A Level
Biology
for OCR
Year 2

Michael Fisher

OXFORD

UNIVERSITY PRESS

OXFORD
UNIVERSITY PRESS

Great Clarendon Street, Oxford, OX2 6DP, United Kingdom

Oxford University Press is a department of the University of Oxford. It furthers the University's objective of excellence in research, scholarship, and education by publishing worldwide. Oxford is a registered trade mark of Oxford University Press in the UK and in certain other countries

British Library Cataloguing in Publication Data
Data available

978 0 19 835776 6

10 9 8 7 6 5 4

Printed by CPI Group (UK) Ltd, Croydon CR0 4YY

Paper used in the production of this book is a natural, recyclable product made from wood grown in sustainable forests. The manufacturing process conforms to the environmental regulations of the country of origin.

Cover: PHILIPPE PSAILA/SCIENCE PHOTO LIBRARY

Artwork by Q2A Media

A Level course structure

This book has been written to support students studying for OCR A Level Biology A. It covers the A Level Year 2 only modules from the specification. The modules covered are shown in the contents list, which also shows you the page numbers for the main topics within each module.

AS exam

A level exam

Year 1 content

1 Development of practical skills in biology
2 Foundations in biology
3 Exchange and transport
4 Biodiversity, evolution, and disease

Year 2 content

5 Communication, homeostasis, and energy
6 Genetics, evolution, and ecosystems

A Level exams will cover content from Year 1 and Year 2 and will be at a higher demand.

This book contains many different features. Each feature is designed to support and develop the skills you will need for your examinations, as well as foster and stimulate your interest in biology.

Worked example

Step-by-step worked solutions.

Common misconception

Common student misunderstandings clarified.

Go further

Familiar concepts in an unfamiliar context.

Maths skills

A focus on maths skills.

Question and model answer

Sample answers to exam style questions.

Practical skill

Support for the practical knowledge requirements of the exam.

Specification references

→ At the beginning of each topic, there are specification references to allow you to monitor your progress.

Key term

Pulls out key terms for quick reference.

Revision tip

Prompts to help you with your understanding and revision.

Synoptic link

These highlight the key areas where topics relate to each other. As you go through your course, knowing how to link different areas of biology together becomes increasingly important. Many exam questions, particularly at A Level, will require you to bring together your knowledge from different areas.

Summary Questions

1 These are short questions at the end of each topic.

2 They test your understanding of the topic and allow you to apply the knowledge and skills you have acquired.

3 The questions are ramped in order of difficulty.

Chapter 13 Practice questions

Practice questions at the end of each chapter including questions that cover practical and maths skills.

1 Which of the following statements is/are true of the functions of the sympathetic nervous system? *(1 mark)*

1 Increases heart rate.

2 Increases the speed at which food moves through the gut.

3 Decreases sweat production.

A 1, 2, and 3 are correct

B Only 1 and 2 are correct

C Only 2 and 3 are correct

D Only 1 is correct

13.1 Coordination

Specification reference: 5.1.1 (a) and (b)

Synoptic link

The remainder of this chapter is concerned with the nervous system. You will learn about two other examples of communication systems in Chapter 14, Hormonal communication, and Chapter 16, Plant responses.

Revision tip: A wonderful reception

The ideas of binding specificity and complementary shapes are relevant to discussions of both enzymes and cell receptors. However, when discussing signalling molecules and cell receptors you should use the term 'binding site'. Reserve the term 'active site' for enzymes.

Multicellular organisms have evolved several specialised systems for responding to internal and external stimuli. These different systems need to be coordinated, which requires cell-to-cell communication (**cell signalling**).

Communication systems

Communication systems involve cells releasing chemicals that stimulate responses in other cells. This is known as cell signalling. **Homeostasis**, which you will read about in Chapter 15, relies on cell signalling between different organs in order to produce coordinated responses.

Cell signalling requires a cell to release a signalling molecule, which binds to **receptors** on a **target cell**.

Examples of communication systems

Communication system	Topic references	How do cells signal to each other?
Nervous system	13.2–13.9	Neurotransmitters (signalling molecules) diffuse from a presynaptic neurone and bind to receptors on postsynaptic neurones.
Animal endocrine system	14.1	Hormones (signalling molecules) are released from glands, travel through blood, and bind to receptors on target cells.
Plant hormonal communication	Chapter 16	Hormones (signalling molecules) diffuse through plant tissue to target cells.

 Go further: Types of cell signalling

Types of cell signalling can be classified based on the destination of the signalling molecule.

Autocrine signalling: the cell that produces the signalling molecule is also the target cell.

Juxtacrine signalling: the target cells are adjacent to the cells producing the signalling molecule.

Paracrine signalling: target cells are nearby the cell that produces the signalling molecule.

Endocrine signalling: target cells are usually not near the producing cells.

1 Suggest, with a reason, the form of cell signalling used by **a** neurones **b** pancreatic β cells.

Summary questions

1 Complete the passage below by placing the most appropriate words or phrases in the gaps.
   In the nervous system, ... act as signalling molecules for neurone-to-neurone communication. The .... neurone is the target cell. In the .... system, hormones act as signalling molecules. *(3 marks)*

2 Describe the role of cell signalling in the nervous system. *(2 marks)*

3 Explain why homeostasis relies on cell signalling. *(2 marks)*

13.2 Neurones

Specification reference: 5.1.3 (b)

The nervous system comprises specialised cells called **neurones**. The electrical impulses transmitted by neurones enable communication between cells in different parts of an organism. Several types of neurone exist.

Neurone structure

A nervous response usually follows this pathway:

Sensory receptor → sensory neurone → relay neurone → motor neurone → effector

▼ **Table 1** *Features of the three types of neurone*

Type of neurone	Function	Key structural features
Sensory	Transmits impulses from receptors to relay neurones in the central nervous system	One **dendron** (carrying the impulse from the receptor to the cell body). One **axon** (carrying the impulse from the cell body to a relay neurone).
Relay	Transmits impulses between neurones	Many clusters of **dendrites**, each leading to a dendron. Each dendron passes to the **central cell body**. A short axon carries impulses from the cell body to many synaptic endings.
Motor	Transmits impulses from a relay neurone to an effector (i.e. muscle or gland)	Dendrites leading to the cell body. One **long axon** (carrying impulses from cell body to effector).

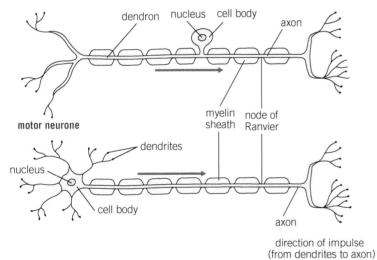

▲ **Figure 1** *Sensory and motor neurones*

Myelinated vs non-myelinated

The axons of motor neurones, many sensory neurones, and some relay neurones are covered in **myelin sheath**, which is a type of fat produced by **Schwann cells**. Myelin speeds up nervous impulses by enabling **saltatory conduction** (see Topic 13.4, Nervous transmission).

Revision tip: Schwann song?
You may be asked to identify myelin sheath and Schwann cells from photographs. Myelin sheath can be seen as a thick band surrounding the axon of a neurone. The Schwann cell will appear as an irregular shape and will be located outside the band of myelin sheath.

Synoptic link
You will learn about action potentials (Topic 13.4, Nervous transmission) and synapses (Topic 13.5, Synapses) later in this chapter.

Revision tip: What a nerve!
Neurones are individual cells. **Nerves** consist of many neurones bundled together.

Revision tip: Cell bodies
The cell body contains a neurone's nucleus and rough ER. This is where new proteins are produced within a neurone.

Revision tip: Is that relay the correct name?
Relay neurones are also known as **interneurones** and **intermediate neurones**.

Summary questions

1 State two structural differences between a sensory neurone and a motor neurone. *(2 marks)*

2 Many relay neurones are not surrounded by myelin sheath. Suggest why. *(2 marks)*

3 Motor neurones can be over one metre long in humans. The cell body of a motor neurone is located at one end of the cell.
   a Suggest one problem that this structural arrangement presents for a motor neurone. *(2 marks)*
   b Suggest which features may be present in the neurone to overcome this problem. *(1 mark)*

13.3 Sensory receptors

Receptors detect changes in the environment and convert the stimulus into an electrical impulse. Stimuli can take many forms, and they can be either internal or external, which has resulted in the evolution of a wide range of receptor structures.

Types of receptor

Receptors are located in sense organs (e.g. the eyes, ears, nose, tongue, and skin). They convert a stimulus to a nervous impulse, which is transmitted along a sensory neurone.

▼ **Table 1** *Types of sensory receptor*

Stimulus	Receptor	Mechanistic detail
Pressure	Pacinian corpuscle	Mechanical pressure applied to the skin opens stretch-mediated sodium ion channels, triggering an action potential (see Topic 13.4, Nervous transmission) in a sensory neurone.
Light	Rods cells (in the retina of the eye)	Light causes a chemical reaction to occur in the rod cells (i.e. the breakdown of rhodopsin), which alters the permeability of the cell membrane to sodium ions.
Chemicals	Taste receptor	Molecules or ions (e.g. sugars, salt, odour molecules) bind to receptors on the receptor cell membranes. This causes a second messenger response, similar to the response produced by hormones (see Topic 14.1, Hormonal communication). cAMP levels rise and alter the permeability of the cell membranes to Na^+ ions. Depolarisation occurs and triggers an action potential.
	Olfactory cells	
Temperature	Thermoreceptors	Specialised sensory neurones. The permeability of their membranes to Na^+ ions changes with temperature.
Sound	Hair cells (in the ear)	Sound waves move cilia on hair cells, which triggers changes in membrane permeability to K^+ ions.

Summary questions

1. Explain why sensory receptors are considered to be transducers.
   (1 mark)

2. State the form of energy converted to electrical impulses by
   a rod cells **b** thermoreceptors **c** Pacinian corpuscles
   d olfactory cells.
   (4 marks)

3. cAMP is generated by the binding of hormones to target cells and, in sensory receptors, in response to certain stimuli. Using the information in Table 1, describe the difference in the action of cAMP in these two types of cell.
   (3 marks)

13.4 Nervous transmission

Specification reference: 5.1.3 (c)

Impulses are transmitted through a neurone by temporary changes in the voltage across the neurone's cell membrane. This change in voltage is called an **action potential**. When no impulses are being transmitted, the voltage across a neurone's membrane is known as the **resting potential**.

Resting potential

The cell membrane of a neurone is **polarised** when not firing impulses (i.e. the extracellular fluid outside the neurone is positively charged, and the cytoplasm in the neurone is negatively charged). The potential difference across the membrane (approximately **−70 mV**) is known as the resting potential. The distribution of two ions (sodium (Na^+) and potassium (K^+)) determines the resting potential.

▼ **Table 1** *How is the resting potential established?*

Membrane protein	How does it produce the resting potential?
Sodium–potassium pump	Three Na^+ ions are pumped out of the neurone (by active transport) for every two K^+ ions pumped in. This sets up an imbalance of positive charge (i.e. outside the neurone is more positive than inside).
K^+ channels	Whereas Na^+ channels are closed at the resting potential, some K^+ channels remain open. This enables some K^+ ions to diffuse out of neurone, down a concentration gradient. Even more positive charge therefore builds up outside the neurone.

Revision tip: In the balance...

The concentration gradient of sodium ions across an axon membrane is steeper than the potassium ion concentration gradient.

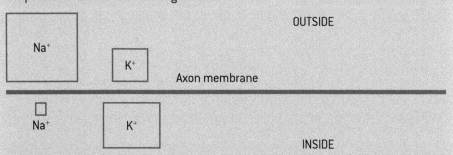

This is why overall the inside of an axon at rest has fewer positive ions than the outside.

Negative ions are also involved in establishing the resting potential, but you are not required to learn about these ions.

Common misconception: Potassium channels

Two forms of potassium ion (K^+) channel proteins are present in neuronal cell membranes.

Some K^+ channels remain open constantly and are not voltage-gated. These channels help establish the resting potential.

Some K^+ channels are voltage-gated. These open during an action potential (stage ④ in Figure 1) in response to an influx of positive charge.

Action potential

The detection of a stimulus by a receptor (see Topic 13.3, Sensory receptors) initiates a change in voltage across a neurone's membrane. The potential difference switches to approximately +40 mV (i.e. the inside becomes more positive than the outside of the neurone). The membrane is **depolarised**. This is known as an action potential.

What happens during an action potential?

1 The **resting potential** (see stage ① in Figure 1).
2 **Na^+ channels open**. Na^+ ions diffuse into the neurone down an **electrochemical gradient**.

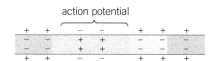

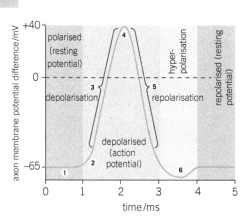

▲ **Figure 1** *Changes in potential difference during an action potential*

Key terms

Resting potential: The potential difference of a neurone membrane at rest (approximately -65 mV).

Action potential: The change in potential difference across a neurone membrane following a stimulus (approximately $+40$ mV).

Synoptic link

The propagation of a nerve impulse is an example of positive feedback, which you will learn about in Topic 15.1, The principles of homeostasis.

3 The initial influx of Na^+ ions causes more voltage-gated Na^+ channels to open (**depolarisation**).

4 Na^+ ions continue to diffuse into the neurone through voltage-gated Na^+ channels until the potential difference reaches $+40$ mV. The voltage-gated **Na^+ channels then close** and voltage-gated **K^+ channels open**.

5 K^+ ions diffuse out of the neurone, reducing the positive charge inside the neurone and **repolarising** the membrane.

6 Voltage-gated K^+ channels close. The membrane becomes **hyperpolarised** (i.e. due to the K^+ ions leaving, the inside of the neurone becomes more negative than it is in the resting state). This is known as a **refractory period** – no more action potentials can occur until the resting potential is restored. Sodium–potassium pumps return the neurone to its resting potential.

Propagation

An action potential is propagated (i.e. spread) along the length of a neurone; this wave of depolarisation is called a **nerve impulse**. The propagation of an action potential involves the following events:

1 Sodium ions enter a neurone and depolarise it.

2 The sodium ions diffuse further along the neurone.

3 The increased positive charge caused by the diffusion of sodium ions opens more (neighbouring) voltage-gated sodium ion channels.

4 The action potential passes along the neurone.

Speed of impulses

Three factors increase the speed of nerve impulses:

- A greater **axon diameter** (which reduces resistance to ion flow).
- A higher **temperature**.
- The presence of **myelin**.

Neurones that are covered in myelin can transmit impulses at a faster rate. Impulses travel faster due to **saltatory conduction** – i.e. the axon membrane can depolarise only at gaps in the myelin (known as **nodes of Ranvier**). The action potential effectively jumps between nodes, which is more efficient than the entire membrane being depolarised.

Key terms

Depolarisation: The change in potential difference across a neurone membrane from negative to positive (e.g. during an action potential).

Repolarisation: The change in potential difference across a neurone membrane from positive to negative (e.g. the restoration of the resting potential).

Hyperpolarisation: The overshoot of a neurone membrane's potential difference following an action potential; it becomes more negative than the resting potential.

Summary questions

1 Complete the following paragraph by adding the most appropriate words or phrases to the gaps. *(4 marks)*
A neurone's resting potential is set up by sodium–potassium pumps, which transfer .... Na^+ ions to the outside of the neurone for every .... K^+ ions transferred inside. During an action potential, the neurone first becomes .... when Na^+ ions diffuse into the axon. The subsequent outward movement of K^+ ions results in ....

2 Describe how the transmission of nerve impulses differs between myelinated and unmyelinated neurones. *(3 marks)*

3 Explain why a neurone will not produce action potentials that vary in magnitude. *(3 marks)*

13.5 Synapses

Once a nervous impulse reaches the end of a neurone, the signal is transmitted to another neurone across a synapse. This is an example of cell signalling.

Synapse structure and transmission

Transmission across a synapse involves the following steps:

1 An action potential passes to the end of the **presynaptic neurone** (presynaptic knob).

2 Voltage-gated **Ca²⁺ ion channels open.**

3 A Ca^{2+} ion influx causes **vesicles** containing neurotransmitters to fuse with the neurone's cell membrane.

4 **Neurotransmitters** are released into the **synaptic cleft** by **exocytosis**.

5 The neurotransmitters **diffuse** across the synaptic cleft and bind to **receptors** on the **postsynaptic neurone's** cell membrane.

6 **Sodium ion channels** are triggered to open, causing depolarisation and an action potential in the postsynaptic neurone (if sufficient neurotransmitters bind and the threshold value is surpassed).

Synapse roles

Synapses play the following roles in the nervous system:

- Ensure impulses travel in only **one direction** (because neurotransmitters are released only from presynaptic neurones, and receptors are found only on postsynaptic neurones).

- One presynaptic neurone can signal to many postsynaptic neurones. This enables signals to be passed to different effectors.

- Enable one sensory neurone to receive signals from several receptors. This provides information about the extent of the stimulus. **Spatial summation** occurs – i.e. sometimes an action potential in a postsynaptic neurone will occur only if several presynaptic neurones release neurotransmitters.

- Enable **temporal summation** – i.e. sometimes a postsynaptic action potential occurs only when several impulses have travelled down a presynaptic neurone. Each impulse releases more neurotransmitters until the threshold depolarisation is surpassed in the postsynaptic neurone.

▲ **Figure 1** *Key features of a synapse (represented here by a cholinergic synapse)*

Revision tip: Excitement or inhibition?

Several neurotransmitters exist. Some (such as acetylcholine) are **excitatory** – i.e. produce action potentials in postsynaptic neurones. Others are **inhibitory** – i.e. they hyperpolarise postsynaptic membranes, reducing the possibility of an action potential.

Revision tip: Cleft and right?

A **synaptic cleft** is specifically the gap between a presynaptic and postsynaptic neurone. The term 'synapse' refers to the combination of the presynaptic membrane, the cleft, and the postsynaptic membrane. You should write that 'neurotransmitters diffuse across the synaptic cleft' (rather than the synapse).

Revision tip: Recycling is important

In most cases, neurotransmitters are reabsorbed back into the presynaptic neurone once they have performed their cell signalling function. **Acetylcholine**, for example, is broken down by an enzyme called **acetylcholinesterase** (found on the postsynaptic membrane). The products (choline and ethanoic acid) are reabsorbed across the presynaptic membrane.

Summary questions

1 Describe how acetylcholine is released from a presynaptic neurone.
(*3 marks*)

2 Use the concept of temporal summation to explain why a weak stimulus may be filtered out and not produce a response from the nervous system.
(*3 marks*)

3 Suggest why it is important for neurotransmitters such as acetylcholine to be recycled.
(*2 marks*)

13.6 Organisation of the nervous system
13.7 Structure and function of the brain

Specification reference: 5.1.5 (h) and 5.1.5 (g)

The mammalian nervous system is organised into neurones with a coordinating role (the central nervous system) and neurones carrying signals to and from the rest of the body (the peripheral nervous system). Neurones within these two regions can be further categorised based on their roles.

Nervous system organisation

The divisions of the nervous system are illustrated in Figure 1.

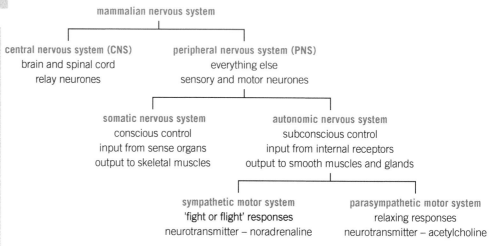

▲ **Figure 1** *Organisation of the mammalian nervous system*

Structure of the brain

The billions of neurones in the brain are organised into regions with different functions.

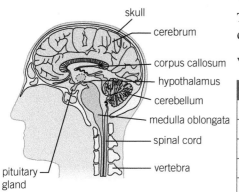

▲ **Figure 2** *The gross structure of the brain*

▼ **Table 1** *The functions of some regions of the brain*

Region of the brain	Function
Cerebrum	Coordinates **voluntary** responses
Cerebellum	Controls **balance** and posture
Medulla oblongata	**Autonomic** functions (e.g. heart rate and breathing rate)
Hypothalamus	**Autonomic** functions (e.g. thermoregulation)
Pituitary gland	Releases **hormones** that control other glands in the body

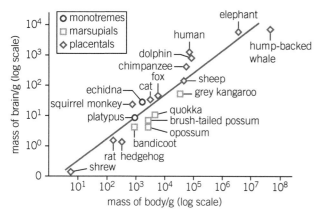

▲ **Figure 3**

13.8 Reflexes

Specification reference: 5.1.5 (i)

A reflex is an involuntary response to a stimulus. They are faster than responses that require conscious thought; such rapid responses can protect the body from danger.

The reflex arc

Rather than using the cerebrum for complex processing, a reflex relies on a simple pathway:

Sensory receptor → sensory neurone → relay neurone → motor neurone → effector

The relay neurones in most reflexes are found in the spinal cord, but some are located in the lower part of the brain.

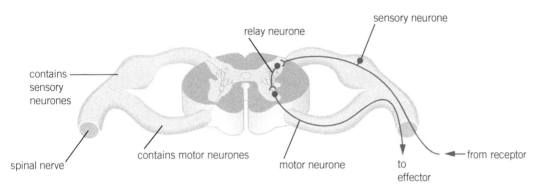

▲ **Figure 1** *A reflex arc*

Examples of reflexes

You need to have an understanding of two reflexes in particular: the knee-jerk reflex and the blinking reflex.

▼ **Table 1** *Two examples of reflexes*

	Knee-jerk reflex	Blinking reflex
Stimulus	Firm tap below the kneecap	Touch on the cornea
Receptor	Stretch receptor in muscle	Touch receptors in the cornea
Location of relay neurone	Spinal cord	Lower brain stem
Effector	Muscles in the upper leg	Muscles in the eyelids
Importance	Maintaining balance	Preventing damage to the eyes

Summary questions

1   State the location of the relay neurones in
    a the knee-jerk reflex b the blinking reflex.   *(2 marks)*

2   Describe and explain the general characteristics of reflex responses
    that aid the survival of organisms.   *(4 marks)*

3   Explain how the typical neural pathway of a reflex maximises its
    effectiveness.   *(3 marks)*

Revision tip: Survival instincts

Reflexes help organisms to survive because they are:

• **very fast** (with only two synapses)

• **innate** (not requiring learning)

• **involuntary** (which frees the brain to process more complex decision-making, if required).

13.9 Voluntary and involuntary muscles
13.10 Sliding filament model
Specification reference: 5.1.5 (I)

A muscle is an example of an **effector**. Impulses transmitted by motor neurones stimulate muscle cells to contract and produce a response.

Types of muscle

Muscles can be either under conscious control (i.e. voluntary, **skeletal muscle**) or under the control of the autonomic nervous system (i.e. involuntary muscle, which can be either **smooth** or **cardiac**).

▼ **Table 1** Types of muscle

Type of muscle:	Skeletal	Cardiac	Smooth (involuntary)
Appearance	Striated (i.e. striped)	Striated (but with fainter striations than skeletal muscle)	Non-striated
Location	Attached to bones via tendons	Heart	Walls of blood vessels, digestive system, and excretory system
Type of contraction	Voluntary (conscious) Fast Short in duration	Involuntary Intermediate speed Intermediate duration	Involuntary Slow Can be long-lasting

Skeletal muscle structure

Muscle cells fuse to form fibres. Each **muscle fibre** contains many **myofibrils**, which are organelles made principally of two proteins: **actin** and **myosin**. Myofibrils are composed of many repeating units called **sarcomeres**.

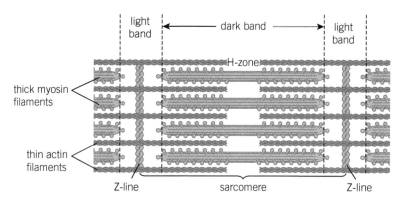

▲ **Figure 1** A sarcomere

Muscle contraction

A muscle contracts when a signal is received from a motor neurone, causing sarcomeres to contract in unison.

Neuromuscular junctions

A **neuromuscular junction** is the synapse between a motor neurone and a muscle fibre. It works using the same principles as a synapse between two neurones: a neurotransmitter (acetylcholine) diffuses across the synaptic cleft and binds to receptors on the sarcolemma, resulting in depolarisation.

A **motor unit** comprises all the muscle fibres supplied by one motor neurone.

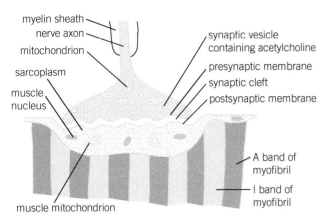

▲ **Figure 2** *A neuromuscular junction*

Sliding filament model

Muscle contraction involves the following steps, which represent the sliding filament model:

1   The sarcolemma is depolarised.

2   The depolarisation spreads through **T-tubules** to **sarcoplasmic reticulum** (specialised smooth ER).

3   **Ca²⁺ ions** are released from the sarcoplasmic reticulum.

4   Ca^{2+} ions bind to **troponin** (a protein that is attached to **actin**).

5   The **troponin changes shape**, which causes **tropomyosin** (another protein) to be moved away from the myosin binding site it had been covering.

6   **Myosin heads** bind to the binding sites on actin. This forms **cross bridges**.

7   Myosin **heads tilt**, thereby moving the actin. This is called the **power stroke**. **ADP is released** from myosin at this stage.

8   **ATP binds** to myosin, causing it to detach from the actin.

9   **ATP is hydrolysed** to ADP, causing the myosin head to resume its original position. The head is free to attach further down the actin. More than 100 power strokes can be performed by each myosin every second.

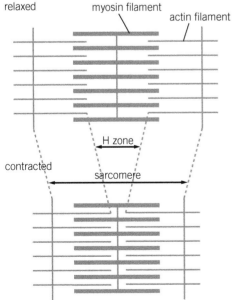

The H zone, sarcomere, and I band all shorten.
The A band is unaltered

▲ **Figure 3** *Changes in a sarcomere during muscle contraction*

Revision tip: Contract changes...

Three changes can be observed when a sarcomere contracts and shortens:

- The light (I) band narrows
- The H zone narrows
- The Z lines move closer.

Revision tip: Energy sources

ATP is supplied to muscle fibres by:

- aerobic respiration (which is why fibres have many mitochondria)
- anaerobic respiration
- **creatine phosphate** (which donates phosphate to ADP to regenerate ATP during intense activity)

Summary questions

1   How will the length of the following sarcomere features change during muscle contraction:
   a   A band
   b   I band
   c   myosin filaments
   d   actin filaments
   e   H zone.   (5 marks)

2   Outline the role of calcium ions in muscle contraction.   (3 marks)

3   Suggest how the ultrastructure of a muscle fibre differs from that of an epithelial cell. Explain the reasons for these differences.   (6 marks)

1 Which of the following statements is/are true of the functions of the sympathetic nervous system? *(1 mark)*

   1 Increases heart rate.

   2 Increases the speed at which food moves through the gut.

   3 Decreases sweat production.

   A 1, 2, and 3 are correct

   B Only 1 and 2 are correct

   C Only 2 and 3 are correct

   D Only 1 is correct

2 What is the function of the cerebellum? *(1 mark)*

   A Coordination of balance and muscular movement.

   B Controls conscious thought processes.

   C Controls body temperature.

   D Regulates heart rate.

3 Which of the following is/are present in the membrane of an acetylcholinergic presynaptic neurone? *(1 mark)*

   1 Calcium ion channel

   2 Acetylcholine receptor

   3 Acetylcholinesterase

   A 1, 2, and 3 are correct

   B Only 1 and 2 are correct

   C Only 2 and 3 are correct

   D Only 1 is correct

4 Which of the following statements is/are true of the role of calcium ions in skeletal muscle contraction? *(1 mark)*

   1 Calcium ions are released into the sarcoplasm.

   2 Calcium ions bind to tropomyosin.

   3 Calcium ions inactivate an enzyme called myosin kinase.

   A 1, 2, and 3 are correct

   B Only 1 and 2 are correct

   C Only 2 and 3 are correct

   D Only 1 is correct

5 Which of the following statements is/are true of the changes in a sarcomere during muscle contraction? *(1 mark)*

   1 Myosin filaments shorten.

   2 The H zone shortens.

   3 I band shortens.

   A 1, 2, and 3 are correct

   B Only 1 and 2 are correct

   C Only 2 and 3 are correct

   D Only 1 is correct

6 Outline the differences between the sympathetic nervous system and the parasympathetic nervous system. *(6 marks)*

14.1 Hormonal communication

Specification reference: 5.1.4 (a) and (b)

You learned about one biological communication system, the nervous system, in the previous chapter. Here you will learn about the endocrine system, which uses chemical messengers called hormones to enable communication between cells.

The endocrine system

Endocrine **glands** secrete chemicals called **hormones** into the circulatory system. Hormones are transported in the **blood** to **target cells**.

Types of hormones

Hormones produce a response in target cells. However, the method by which they produce a response depends on the type of hormone.

Type of hormone	Chemistry	How does it affect target cells?	Examples
Steroid	Lipid-soluble	Diffuses through the cell surface membrane Binds to a receptor (in cytoplasm or nucleus) Promotes or inhibits transcription (of a particular gene)	Testosterone Oestrogen Glucocorticoids Mineralocorticoids
Non-steroid	Hydrophilic (e.g. polypeptides or glycoproteins)	Binds to a receptor on the cell surface membrane Activates a second messenger (e.g. cyclic AMP) A cascade of intracellular reactions activates a transcription factor (for a particular gene)	Glucagon Insulin Adrenaline

The adrenal glands

The adrenal glands produce both steroid hormones (from the cortex region) and non-steroid hormones (from the medulla).

Region of adrenal glands	Hormone	Function
Cortex	Glucocorticoids	Regulates carbohydrate metabolism
	Mineralocorticoids	Regulates salt and water concentrations
	Androgens (e.g. testosterone)	Various functions (e.g. regulation of muscle mass, development of sexual characteristics)
Medulla	Adrenaline	Increases heart rate and blood glucose concentration (by stimulating glycogenolysis)
	Noradrenaline	Works in concert with adrenaline by increasing heart rate, widening airways, and increasing blood pressure

Summary questions

1   Define **a** a target cell **b** a second messenger. *(3 marks)*

2   Outline the similarities and differences between the modes of action of steroid and non-steroid hormones. *(4 marks)*

3   Suggest how the adrenal glands respond to physiological changes in an athlete running a marathon. *(5 marks)*

14.2 The pancreas

The pancreas is an organ that has both **endocrine** functions (i.e. hormone production) and **exocrine** functions (i.e. enzyme production).

Pancreatic functions

Role of the pancreas	What is produced?	Where is it produced?
Exocrine gland	Amylase	These digestive enzymes are secreted from **exocrine tissue** (which constitutes most of the pancreas) and are released into the **pancreatic duct**, which leads to the duodenum (the upper small intestine).
	Proteases	
	Lipases	
Endocrine gland	Insulin	β (beta) cells (in the islets of Langerhans).
	Glucagon	α (alpha) cells (in the islets of Langerhans).

Pancreatic structure

The pancreas contains two types of cell (the **islets of Langerhans** and exocrine cell clusters called **acini**), which can be distinguished through staining.

Structure	How can you identify them?
Islets of Langerhans (endocrine)	A large cluster of cells that is usually stained **blue/lilac** in photographs
Acini (exocrine)	Small clusters of cells stained **dark pink/purple**

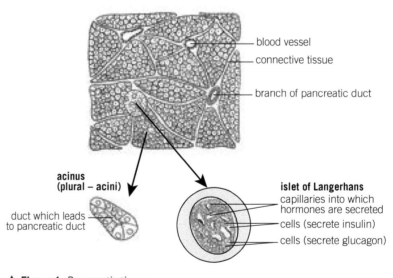

▲ **Figure 1** *Pancreatic tissue*

Summary questions

1 Describe how cells in the islets of Langerhans can be distinguished from pancreatic acini cells. (*2 marks*)

2 Explain the difference between exocrine and endocrine glands, using the pancreas as an example. (*4 marks*)

3 Suggest how transmembrane transport of ions and molecules across the cell membranes of pancreatic α and β cells will differ. (*5 marks*)

14.3 Regulation of blood glucose concentration

Specification reference: 5.1.4 (d)

The regulation of blood glucose is coordinated by endocrine cells in the pancreas and is an example of homeostasis.

How does the pancreas regulate blood glucose concentration?

	When blood glucose needs to be increased	When blood glucose needs to be decreased
What is secreted?	Glucagon (from α cells)	Insulin (from β cells)
Where does it have an effect?	Binds to receptors on liver cells	Binds to receptors on liver and muscle cells
What effects are produced?	Less glucose taken up by cells More fatty acids are used in respiration Glycogen converted to glucose (**glycogenolysis**) Amino acids and fats converted to glucose (**gluconeogenesis**)	More glucose absorbed by cells Glucose is converted to fats More glucose used in respiration Glucose converted to glycogen (**glycogenesis**) Inhibits glucagon release

Insulin secretion

Insulin is secreted from β cells using the following mechanism:

1 Blood glucose concentration is high. **2 Glucose enters** a β cell through a **transporter** protein in the cell surface membrane. **3** Glucose is metabolised to produce **ATP. 4** ATP binds to and closes **K⁺ ion channels** in the cell surface membrane. **5** K^+ ions build up in the cell and increase the positive charge in the cell (**depolarisation**). **6** Voltage-gated **Ca²⁺ ion channels** open. **7** Ca^{2+} ions diffuse into the cell and trigger secretory **vesicles** to release **insulin** from the cell via **exocytosis**.

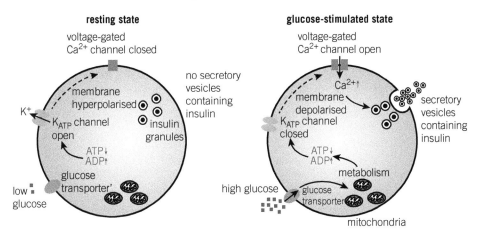

▲ **Figure 1** *The mechanism of insulin release in beta cells*

Revision tip: (Negative) feedback is welcome

Insulin and glucagon have antagonistic effects. The lowering of blood glucose by insulin and the raising of blood glucose by glucagon are examples of **negative feedback**.

Synoptic link

You learned about the structure of glucose in Topic 3.3, Carbohydrates. In Topic 15.1, The principles of homeostasis, we will discuss the concept of negative feedback.

Revision tip: G is for... ?

Many of the terms in this topic begin with 'g'. Understanding the following is important:

'lysis' = splitting
'genesis' = creation/formation
'neo' = new

Therefore:
glycogenesis = the formation of glycogen
glycogenolysis = the splitting of glycogen
gluconeogenesis = the formation of new glucose

Summary questions

1 Describe the role of ATP in the release of insulin from beta cells in the pancreas.
(*2 marks*)

2 The set point for blood glucose concentration in humans is approximately 90 mg 100 cm⁻³. How would this value be expressed in g cm⁻³?
(*1 mark*)

3 Suggest why liver cells have specialised receptors for glucagon.
(*3 marks*)

14.4 Diabetes and its control

Specification reference: 5.1.4 (e) and (f)

Diabetes mellitus is a disease in which the homeostatic control of blood glucose concentration is lost. Untreated, the level of glucose in the blood of a diabetic patient is liable to be too high.

Types of diabetes

Characteristic	Type 1 (insulin dependent)	Type 2 (insulin independent)
Insulin production	Little or none	Often reduced
Cause	Usually an autoimmune response (β cell destruction, which stops insulin production)	Effector cells lose responsiveness to insulin
Genes vs environment?	Genetic	Genetic and environmental (e.g. diet and activity levels) factors influence the risk of developing the disease
Age at onset	Childhood (juvenile-onset)	Adulthood (late-onset)
Speed of onset	Quick	Slow
Usual treatments	Insulin injections	Dietary control of carbohydrate intake

Go further: Maturity-onset diabetes of the young

Forms of diabetes mellitus exist other than type 1 and 2, although they are rare. Maturity-onset diabetes of the young (MODY) is a hereditary condition that constitutes 2% of diabetes mellitus cases. It shows autosomal dominant inheritance (you can remind yourself of inheritance patterns by reading Topic 20.2, Monogenic inheritance).

MODY tends to be diagnosed in people below 30 years of age. It is usually characterised by mild hyperglycaemia (high blood glucose concentration). MODY can be controlled in the early stages by planning meals and may not require insulin injections. Patients are not insulin resistant; the problem lies with glucose metabolism or insulin secretion. Several different forms of MODY have been identified, each caused by a genetic variant at a single gene locus.

1 Suggest why MODY is sometimes referred to as 'monogenic diabetes'.

2 What is the range of probability that two parents with MODY will have a child with the condition?

3 MODY patients with heterozygous genotypes can often manage their condition through careful dietary planning. Suggest why MODY patients with homozygous genotypes are likely to require insulin injections.

Synoptic link

In Topic 21.4, Genetic engineering, you will learn more about the process of genetically modifying bacteria to produce insulin and other proteins. You learned about stem cells in Topic 6.5, Stem cells.

Question and model answer: Insulin sources

Q. Describe how the sources of insulin for type 1 diabetes treatment are changing, and explain the benefits of these changes.

A. For many years, insulin was obtained from the pancreas of other animals (e.g. pigs).

The production of human insulin was made possible by using genetically modified (GM) bacteria. This method has some advantages:

- It's cheaper than extracting non-human insulin from other animals.

- The rate of production is higher.

> When answering these type of questions, ensure you link each description to at least one explanation.

More recently, some patients have been injected with β cells. Stem cell therapy is being researched as an alternative treatment. This has advantages over GM insulin production and β cell transplant:

- Patients would no longer require insulin injections.

- There would be a low risk of rejection.

Summary questions

1 Why is type 1 diabetes sometimes called insulin-dependent and type 2 called insulin-independent? *(2 marks)*

2 Study Figure 1. Describe and explain the differences in the responses of diabetics and non-diabetics following the consumption of glucose. *(3 marks)*

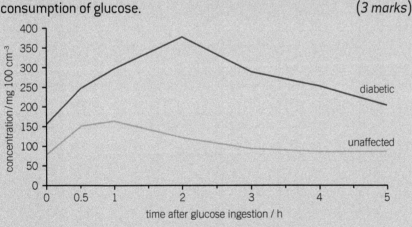

▲ Figure 1

3 Suggest why the prevalence of type 2 diabetes is likely to have a greater impact on future populations than the prevalence of type 1 diabetes. *(5 marks)*

14.5 Coordinated responses

Specification reference: 5.1.5 (j)

The nervous and endocrine (hormonal) systems often work together in response to stimuli. The **fight or flight** response in mammals is an example of such a coordinated response.

Endocrine and nervous system responses

▼ **Table 1** *A comparison of the endocrine and nervous systems*

	Endocrine (hormonal) system	Nervous system
What is transmitted?	Chemicals (hormones) in the blood	Electrical impulses along neurones
How quick is the communication?	Relatively slow	Rapid
For how long do the effects last?	Long-lasting and can be permanent	Short-lived and temporary

Fight or flight response

Mammals react to danger through the fight or flight response. This response is coordinated by the **sympathetic** branch of the **autonomic nervous system** (and is controlled by the **hypothalamus**). However, hormonal communication also plays a role; the sympathetic nervous system stimulates the release of many hormones, including **adrenaline** (from the adrenal medulla) and **cortisol** (a stress hormone produced in the adrenal cortex).

▼ **Table 2** *Fight or flight responses*

Physical response		How does this benefit the mammal?
Increase in…	Heart rate	Oxygen and glucose are circulated faster
	Blood glucose concentration	Respiration rate in cells is raised
	Pupil dilation	Improves vision
Decrease in…	Blood flow to skin surface	More blood can be diverted to skeletal muscle, the brain, and the heart
	Digestion rate	
	Concentration on small tasks	The brain focuses on the immediate threat

How does adrenaline produce its effects?

Adrenaline is a non-steroid hydrophilic hormone. You read about adrenaline and the action of non-steroid hormones in Topic 14.1, Hormonal communication. Adrenaline produces effects in the target cell using the following mechanism:

- It binds to a cell surface **receptor**.
- It activates **adenylyl cyclase** (which is an enzyme).
- Adenylyl cyclase converts ATP to cyclic AMP (**cAMP**) (a second messenger).
- cAMP activates protein kinases, which activate other enzymes (this is known as a **cascade effect**).

Synoptic link

Chapter 13 focused on the nervous system, and you learned about the structure and function of the adrenal glands in Topic 14.1, Hormonal communication.

Summary questions

1 Describe the role of adenylyl cyclase in a liver cell. *(2 marks)*

2 Explain why the fight or flight response is considered a coordinated response (i.e. controlled by both the nervous system and the endocrine system). *(2 marks)*

3 Suggest why a mammal responding to danger may experience a loss of hearing. *(2 marks)*

14.6 Controlling heart rate

Specification reference: 5.1.5 (k)

Heart rate is controlled by the autonomic nervous system and, in times of stress, hormones.

How is heart rate controlled?

Changes in heart rate are controlled by the cardiovascular centre in a region of the brain called the **medulla oblongata**.

▼ **Table 1** *How changes to heart rate are initiated by the medulla oblongata*

	To increase heart rate	To decrease heart rate
Which branch of the autonomic nervous system is used?	Sympathetic	Parasympathetic
Which nerve transmits the impulse from the medulla oblongata?	Accelerator nerve	Vagus nerve
What effect is produced in the heart?	Increases the rate at which the SAN generates impulses	Decreases the rate at which the SAN generates impulses

Role of receptors

Two different types of receptor pass information to the medulla oblongata: chemoreceptors and pressure receptors (baroreceptors).

▼ **Table 2** *The roles of receptors in the regulation of heart rate*

	Chemoreceptors	Baroreceptors
Where are they located?	Aorta, carotid artery (located in the neck), and the medulla oblongata	Aorta, carotid artery, and vena cava
What do they detect?	Changes in pH	Changes in blood pressure
What responses are produced?	High carbon dioxide concentration in the blood results in low pH, which triggers an increase in heart rate. Low carbon dioxide concentration means a higher pH, which causes a decrease in heart rate.	Heart rate is increased when blood pressure is too low. Heart rate is decreased when blood pressure is higher than normal.

Role of adrenaline

Adrenaline and noradrenaline **increase heart rate** by binding to cardiac cells and increasing the frequency of impulses generated by the SAN.

Synoptic link

You learned about the structure of the heart and the role of the SAN in controlling the cardiac cycle in Topic 8.5, The heart.

Revision tip: Many medullas

A **medulla** is the inner region of a tissue or organ. Other organs, such as the kidney, contain a medulla. Try to refer to the medulla oblongata by its full name to avoid confusion.

Revision tip: Regulation rather than initiation

The cardiac muscle is myogenic, which means its contractions are initiated within the heart tissue rather than by an external stimulus. The accelerator and vagus nerves do not make the heart muscle contract, they regulate the rate at which the SAN fires.

Synoptic link

Topics 14.1, Hormonal communication, and 14.5, Coordinated responses, examine how adrenaline increases heart rate.

Maths skill: Student's *t*-test

A *t*-test is used to assess whether a significant difference exists between two sets of data. Heart rate data lend themselves to this form of analysis. For example, we could compare heart rate before and after exercise, or compare the average resting heart rate of smokers and non-smokers.

You can remind yourself of how to calculate *t* by revisiting Topic 10.6, Representing variation graphically.

Here are some points to remember when using a *t*-test:

- Two data sets are required.
- Mean values are compared.
- The probability value at which biologists tend to assign significance is 0.05. This indicates that there is only a 5% probability that the difference between two data sets is due to chance.

Summary questions

1 Describe the role of receptors in regulating heart rate. *(3 marks)*

2 Explain why heart rate must be altered in response to increased physical activity. *(3 marks)*

3 Imagine that the vagus nerve of an organism is cut. Suggest what would happen if the organism's blood pressure increases above its normal value. *(3 marks)*

1 Which of the following processes is a result / are results of the
   secretion of glucagon from the pancreas?   (*1 mark*)

   1 Glycogenolysis

   2 Gluconeogenesis

   3 Glycogenesis

   A 1, 2, and 3 are correct

   B Only 1 and 2 are correct

   C Only 2 and 3 are correct

   D Only 1 is correct

2 Which of the following characteristics is generally true of
   type 1 diabetes?   (*1 mark*)

   A Slow speed of onset.

   B Late onset.

   C Caused by an autoimmune response.

   D Insulin-independent.

3 The set point for blood glucose concentration in humans is
   approximately $9 \times 10^{-4}\,\mathrm{g\,cm^{-3}}$. How would this value be expressed
   in $\mathrm{mg\,100\,cm^{-3}}$?   (*1 mark*)

   A 0.9

   B 9

   C 90

   D 900

4 Describe how heart rate is controlled by both nervous and
   hormonal mechanisms.   (*6 marks*)

5

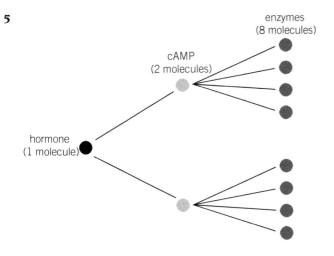

   Use the diagram to describe and explain how adrenaline can increase
   blood glucose concentration.   (*5 marks*)

6 Copy and complete the table to show the differences between type 1 and
   type 2 diabetes.

Trait	Type 1	Type 2
Treatment		
Extent of genetic influence		
Is insulin produced?		

                                                               (*3 marks*)

15.1 The principles of homeostasis

Specification reference: 5.1.1(c)

The maintenance and control of an organism's internal conditions within narrow limits is known as **homeostasis**.

Negative feedback

Homeostasis relies on control systems that detect changes in internal conditions and produce the required responses to reverse these changes. The corrective mechanism that allows only small fluctuations around a set point is known as **negative feedback**.

A control system must have the following features:

- A **set point**: this represents the desired value around which the negative feedback mechanism operates. Physiological factors tend to vary over a small range either side of the set point. This represents the **normal range** of that factor.
- **Receptors**: they detect stimuli and deviations from the set point.
- **Controller** (communication pathway): this coordinates the information from the receptors and sends instructions to the effectors. The nervous system and the hormonal system tend to act as controllers.
- **Effectors**: they produce the changes required to return the system to the set point.
- **Feedback loop**: the return to the set point creates a feedback loop.

> ### Key term
>
> **Homeostasis:** The maintenance of stable conditions (within narrow limits) inside the body.

> ### Key term
>
> **Negative feedback:** The mechanism controlling homeostasis; a change in a parameter leads to the reversal of the change.

factor rises above normal level → body detects the change and initiates a corrective mechanism → factor returns to normal level → factor at normal level

factor falls below normal level → body detects the change and initiates a corrective mechanism → factor returns to normal level → factor at normal level

▲ **Figure 1** *Negative feedback*

 Worked example: Applying terms to an example of homeostasis

Apply the terms 'set point', 'normal range', 'receptors', 'controller', and 'effectors' to the homeostatic control of water potential in the blood (see Topic 15.5, The structure and function of the mammalian kidney).

Set point = 285 mOsm kg^{-1}

Normal range = 275–295 mOsm kg^{-1}

Receptors = osmoreceptors (in the hypothalamus)

Controller = hypothalamus (secreting ADH)

Effectors = cells of the collecting duct (in the kidneys)

Positive feedback

Negative feedback reverses a change in conditions, whereas positive feedback increases the original change. Negative feedback **inhibits** the original stimulus, but positive feedback **enhances** the original stimulus.

Positive feedback is not involved in homeostasis. Examples include:

- the attraction of platelets to a site of blood clotting
- the generation of action potentials in neurones (see Topic 13.4, Nervous transmission)
- childbirth (the hormone oxytocin is released when the baby's head pushes against the cervix; oxytocin stimulates uterine contractions, which causes more of the hormone to be produced).

Summary questions

1 In the context of a control system, state what is meant by the terms **a** set point **b** normal range. *(2 marks)*

2 Describe the positive feedback mechanism that controls uterine contractions during labour, and suggest how positive feedback is beneficial during this process. *(4 marks)*

3 Explain why homeostasis relies on negative feedback and not positive feedback. *(3 marks)*

15.2 Thermoregulation in ectotherms
15.3 Thermoregulation in endotherms

Specification reference: 5.1.1(d)

You learned about the general principles of homeostasis in the previous topic. One example of homeostasis is **thermoregulation** – the control of body temperature.

In ectotherms

Environmental temperature rather than metabolism dictates body temperature in **ectotherms** (animals such as fish, amphibians, and reptiles). Ectotherms show behavioural responses to regulate their internal temperature.

▼ **Table 1** *Ectotherm behaviours*

Behaviour when too cold	Behaviour when too hot
Basking (exposing their bodies to the Sun)	Finding shade or burrowing
Change body shape (e.g. increase surface area to gain heat in hot weather; decrease surface area to retain heat in cold weather)	
Pressing body against warm ground (to gain heat through conduction)	Pressing body against cool stones (to lose heat through conduction)

In endotherms

Endotherms (mammals and birds) use internal mechanisms to control body temperature.

The homeostatic control of body temperature relies on the following components:

Receptors: peripheral temperature receptors (in skin), temperature-sensitive neurones in the hypothalamus (to monitor core temperature).

Controller: hypothalamus

Effectors: e.g. sweat glands, erector muscles controlling hair follicles, skeletal muscle, sphincter muscles controlling vasodilation and vasoconstriction in arterioles.

▼ **Table 2** *Endotherm responses to internal temperature changes – these are negative feedback responses controlled by the hypothalamus*

To warm up	Explanation	To cool down	Explanation
Less sweat	Less heat lost through evaporation of sweat	More sweat	Sweat evaporation requires heat from the blood, producing a cooling effect
Hairs raised	A layer of insulating air is trapped	Hairs lie flat on the skin	More heat can be lost through radiation
Vasoconstriction	Less blood flows through capillaries near the skin surface; less heat is radiated from the body	Vasodilation	More blood flows through capillaries near the skin surface; more heat is radiated from the body
High metabolic rate in liver cells	Respiration generates more heat	Low metabolic rate in liver cells	Respiration generates less heat
Skeletal muscles contract spontaneously (shivering)	Heat generated from respiration	No spontaneous contractions	No additional heat generated from respiration in muscles

Revision tip: Ectotherm physiological responses and endotherm behaviours

In general, ectotherms use behaviours to warm up or cool down, whereas endotherms rely on physiology to a greater extent. However, some ectotherms have evolved physiological adaptations (e.g. dark pigments to absorb radiation, and the ability to alter metabolic rate), and endotherms tend to show the same range of behavioural responses exhibited by ectotherms.

Summary questions

1 Name four structures located in the skin that are used by endotherms in thermoregulation. *(4 marks)*

2 Discuss the advantages and disadvantages of ectothermy. *(4 marks)*

3 Suggest why the food intake of an endothermic species is likely to increase during the winter compared to the summer months. *(3 marks)*

15.4 Excretion, homeostasis, and the liver

Specification reference: 5.1.2(a) and (b)(i–ii)

Excretion is the removal of waste products from the body. The liver is an organ that plays an important role in excretion, as well as having a range of other functions.

Excretion

Examples of substances that need to be excreted include:

- Carbon dioxide (see Topics 7.1 to 7.4).
- Urea ($CO(NH_2)_2$), which is produced in the liver from ammonia and carbon dioxide).
- Bile pigments (which form when haemoglobin is broken down by **Kupffer cells** – specialised macrophages in the liver).

The liver

Structure

The liver is divided into units called **lobules**. Each lobule contains liver cells (**hepatocytes**), which border two types of channel: **sinusoids** (vessels in which blood from the hepatic artery and hepatic portal vein mix) and **bile canaliculi** (which drain bile into the gall bladder).

Functions of the liver

Function	Description
Deamination	An amino acid can be converted to ammonia (NH_3) and a keto acid. Ammonia is then converted to urea in the **ornithine cycle** ($2NH_3 + CO_2 \rightarrow CO(NH_2)_2 + H_2O$).
Transamination	The conversion of one amino acid to another.
Glycogen storage	Glucose is converted to glycogen in hepatocytes; glycogen stores are hydrolysed to glucose when required (see Topic 14.3).
Detoxification	e.g. hydrogen peroxide (a metabolic waste product) is converted to water and oxygen by catalase. e.g. ethanol is converted to ethanal by alcohol dehydrogenase.

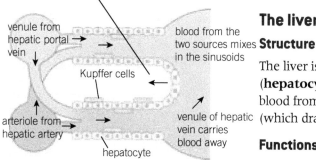

bile canaliculus – the hepatocytes secrete bile into these canaliculi and from there it drains into the gall bladder to be stored

venule from hepatic portal vein

blood from the two sources mixes in the sinusoids

Kupffer cells

arteriole from hepatic artery

venule of hepatic vein carries blood away

hepatocyte

▲ **Figure 1** *The structure of a liver lobule*

Revision tip: Blood ties

The liver is supplied by blood from two vessels:

1 The **hepatic artery** supplies oxygenated blood from the heart.

2 The **hepatic portal vein** supplies blood carrying glucose, amino acids, and fats from the small intestine.

However, the **hepatic vein** transports deoxygenated blood *from the liver*.

Synoptic link

You may be required to examine and draw liver tissue under a microscope. Remind yourself of some basic microscopy skills by reading Topic 2.1, Microscopy.

Common misconception: Excretion, egestion, secretion

Excretion is the removal of metabolic waste from the body. Do not confuse this with **egestion** (i.e. defecation – the removal of undigested material from the body) or **secretion** (i.e. the release of a substance for a particular function – e.g. sweat secretion).

Summary questions

1 State two examples of substances that the liver removes in its role as an organ of detoxification. *(2 marks)*

2 Explain the importance of transamination. *(2 marks)*

3 Describe the role of enzymes in aspects of liver function. *(5 marks)*

15.5 The structure and function of the mammalian kidney

Specification reference: 5.1.2(c)i–iii, (d), (e), and (f)

Mammalian kidneys perform two homeostatic functions: **excretion** (see Topic 15.4, Excretion, homeostasis, and the liver) and **osmoregulation** (the control of water and ion concentrations in an organism).

Kidney structure

The two kidneys in vertebrates are supplied with blood by the **renal artery**, and a **renal vein** drains each organ. **Urine** produced by the kidneys passes through the **ureter** to the **bladder**.

Each kidney has three distinct regions:

- The **cortex** (the outer region).
- The **medulla** (the inner region).
- The **renal pelvis** (the most central region, which leads to the ureter).

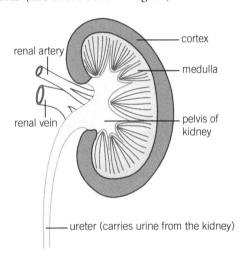

▲ **Figure 1** *The gross structure of a kidney*

The functional unit of the kidney is called a **nephron**; each kidney contains approximately one million nephrons.

Synoptic link

You may be required to examine and draw kidney tissue under a microscope. Remind yourself of some basic microscopy skills by reading Topic 2.1, Microscopy.

Revision tip: Three filters

Three different filters prevent cells and large molecules from leaving the blood during ultrafiltration:

Narrow gaps (**fenestrations**) between endothelium cells in capillaries.

Podocytes (epithelial cells of the capsule that have finger-like projections, which form filtration slits).

Basement membrane (a mesh of collagen and glycoprotein around the glomerulus).

Functions of regions of the nephron

Region	Function	How does it work?
Bowman's capsule	Ultrafiltration	Water and other small molecules are forced from the blood (a network of capillaries called the **glomerulus**) into the Bowman's capsule. Large molecules and blood cells remain in the glomerulus. High **hydrostatic pressure** is created in the glomerular capillaries by the **afferent** (incoming) arteriole being wider than the **efferent** (outgoing) arteriole. The fluid that is filtered into the capsule is called glomerular **filtrate**.
Proximal convoluted tubule (PCT)	Selective reabsorption	All glucose, amino acids, vitamins, and hormones are reabsorbed into the blood. 85% of water and NaCl is reabsorbed. The cells of the PCT wall are specialised for reabsorption in the following ways: • **Microvilli** to increase surface area for reabsorption • Plasma membranes have many **pumps** and **transporter proteins** for active transport and facilitated diffusion • Many **mitochondria** to produce ATP for active transport. Toxic compounds (e.g. urea), excess water, and some ions remain in the filtrate.
Loop of Henle	Establishing a water potential gradient	The filtrate and blood are **isotonic** at the beginning of the loop of Henle (i.e. no water potential gradient exists). By the end of the loop of Henle, the water potential in the filtrate is higher than in the medulla. This enables additional water to be reabsorbed from the collecting duct down a **water potential gradient**, if necessary. The **descending limb** is permeable to water. However, the **ascending limb** is impermeable to water, but Na^+ and Cl^- ions are actively transported from the filtrate into the medulla, which raises the water potential of the filtrate.
Distal convoluted tubule	Additional ion reabsorption	Additional active transport and reabsorption of ions can occur here.
Collecting duct	Determination of urine concentration and volume	Water can be reabsorbed, depending on the body's needs. The permeability of the collecting duct (and therefore the volume of water that is reabsorbed) is controlled by ADH (see below).

Osmoregulation

Water potential is controlled through a negative feedback mechanism. A decrease in blood water potential below the set point causes the following response:

• **Osmoreceptors** in the **hypothalamus** detect the low water potential.
• Neurosecretory cells (specialised nerve cells) in the hypothalamus are stimulated to release **ADH** from the **posterior pituitary gland**.

- ADH travels through the blood to the kidneys.
- ADH binds to receptors on the cell surface membrane of cells in the **collecting duct wall**.
- The concentration of cyclic AMP (**cAMP**) in these cells is increased.
- cAMP acts as a second messenger and causes **aquaporins** to be inserted into the membrane of cells in the collecting duct wall.
- Aquaporins are membrane-spanning channel proteins that increase the permeability of the collecting duct wall. They allow water to diffuse through but prevent the passage of ions.
- More water is reabsorbed from the collecting duct by osmosis.
- Urine with a high solute concentration is produced.
- The water potential of the blood is increased.

The same system responds if blood water potential rises above the set point. Less ADH is produced, and less water is reabsorbed into the blood from the collecting duct. A large volume of dilute urine is produced.

Revision tip: Urea vs. urine

Make sure you know the difference between urea and urine. Urea is the waste molecule $CO(NH_2)_2$, whereas urine is the fluid produced by the kidneys that contains dissolved urea.

Go further: Diuretics – treating, cheating, and energy-depleting?

A **diuretic** is a substance that increases the volume of urine produced by an individual. There are many examples, which produce their effects through a variety of different mechanisms.

Diuretics have many medical uses. High blood pressure (hypertension) can be treated with thiazide diuretics. Thiazides target cells in the distal convoluted tubule wall. They inhibit sodium–chloride ion cotransport, which leads to more water being excreted in the urine.

Some athletes use diuretics to cheat by invalidating their drug tests. By raising their urine volume, athletes can dilute performance-enhancing drugs and their metabolites. The diuretics act as masking agents.

Alcohol acts as a diuretic by inhibiting ADH production. Hangovers are not only a result of alcohol toxicity; they are also caused by the effect on brain cells of blood with a high solute concentration.

1 Suggest why thiazide diuretics result in a higher volume of urine being excreted.

2 Explain how drinking plenty of water after heavy alcohol consumption could limit the effects of alcohol on brain cells.

Uses of excretory products

The excretory products in urine can be analysed to make medical diagnoses.

Excretory product	What is diagnosed?	Details
Glucose	Diabetes	Glucose should be reabsorbed in the PCT of nephrons. The presence of glucose may indicate diabetes.
Human chorionic gonadotrophin (hCG)	Pregnancy	**Monoclonal antibodies** detect the presence of hCG (which is produced during pregnancy) in urine.
Anabolic steroids (and other drugs)	The use of banned drugs in sport	Urine samples are tested using gas chromatography and mass spectrometry.

Kidney failure

Kidney failure is either acute or chronic. Kidneys can fail for the following reasons: bacterial infection, kidney stones, uncontrolled diabetes (type 1 or 2), hypertension, and certain inherited diseases.

Treatments

Treatment	Description	Advantages	Disadvantages
Transplant	Surgery to replace a failed kidney with a donor's kidney	If successful, removes the need for dialysis treatment	Possible transplant rejection (therefore patient requires immunosuppressant drugs)
Haemodialysis	A patient's blood is filtered through a dialysis machine	Daily dialysis is not required	Patients must spend hours in hospitals each week
Peritoneal dialysis	A patient's abdominal membrane (peritoneum) is use as a dialysis membrane	It can be done at home, and patients are able to walk and work during dialysis	Must be performed daily and requires an initial implantation of a tube

Revision tip: Acute vs. chronic

Remember that an **acute** condition has a quick onset but often a fast recovery time. A **chronic** condition tends to develop more slowly but usually lasts a long time.

Summary questions

1   Outline the likely differences in composition between blood in the renal artery and blood in the renal vein. *(2 marks)*

2   The cells lining the proximal convoluted tubule use endocytosis and exocytosis, in addition to active transport and facilitated diffusion, to move molecules across their membranes. Suggest why endocytosis and exocytosis are used. *(3 marks)*

3   Evaluate the costs and benefits of dialysis and transplant surgery as treatments for kidney failure. *(6 marks)*

1   Which of the following statements is an example / are examples
    of positive feedback?   (*1 mark*)

    1   The mechanism by which platelets are attracted to a site of blood clotting.

    2   The interaction between oxytocin and contraction strength during
        childbirth.

    3   The development of hypothermia.

    A   1, 2, and 3 are correct

    B   Only 1 and 2 are correct

    C   Only 2 and 3 are correct

    D   Only 1 is correct

2   Which of the following is a response to the body's temperature
    rising above its set point?   (*1 mark*)

    A   Vasoconstriction.

    B   Hairs lie flat.

    C   Increased muscle contraction within some tissues.

    D   Decreased sweat production.

3   Which of the following is an adaptation / are adaptations in a
    nephron to enable ultrafiltration?   (*1 mark*)

    1   Cells in capillary walls are separated by narrow gaps.

    2   The endothelium cells of the capillaries have finger-like projections
        called podocytes.

    3   Basement membranes allow large proteins to pass through into the
        nephron filtrate.

    A   1, 2, and 3 are correct

    B   Only 1 and 2 are correct

    C   Only 2 and 3 are correct

    D   Only 1 is correct

4   In which region of a nephron does the majority of selective
    reabsorption occur?   (*1 mark*)

    A   Proximal convoluted tubule.   C   Distal convoluted tubule.

    B   Loop of Henle.   D   Collecting duct.

5   Which of the following is / are true of antidiuretic hormone?   (*1 mark*)

    1   It is released from the posterior pituitary gland.

    2   It binds to cells in the collecting duct wall.

    3   It results in aquaporins being inserted into the membrane of cells in the
        collecting duct wall.

    A   1, 2, and 3 are correct   C   Only 2 and 3 are correct

    B   Only 1 and 2 are correct   D   Only 1 is correct

6   Which of the following statements is true of peritoneal dialysis?   (*1 mark*)

    A   It is usually performed three times per week.

    B   A patient's blood is passed through a dialysis machine and
        returned to their body.

    C   It must be performed every day.

    D   It must be carried out by medical professionals.

16.1 Plant hormones and growth in plants

Specification reference: 5.1.5(b), (c), (d) and (e)

Synoptic link

You learned about mammalian hormones in Chapter 14, Hormonal communication.

Like animals, plants produce hormones. These chemicals enable cell signalling and communication between different parts of a plant. Hormones work in concert to create coordinated responses to environmental conditions.

Roles of hormones

Auxins and **gibberellins** are two families of hormones responsible for a range of functions.

Hormone	Role	How does it operate?	Experimental evidence
Auxins	Cell elongation	Increases cell wall stretchiness, enabling water absorption and cell expansion	Auxin application causes pH to decrease. Enzymes that increase cell wall flexibility work best in acidic conditions.
	Apical dominance	Promotes growth of the main shoot and inhibits lateral shoots	Lateral shoots grow faster when the apex (i.e. tip) of the main shoot (the site of auxin production) is removed.
	Root growth	Low auxin concentrations promote root elongation	Application of small quantities of auxin stimulates root growth; high concentrations inhibit growth.
Gibberellins	Stem elongation	Stimulates cell elongation and division	Gibberellin concentrations and plant height show a positive correlation.
	Seed germination	Activates genes for amylase and protease enzymes, which break down food stores in seeds	Mutant seeds that lack gibberellin genes and seeds exposed to gibberellin inhibitors do not germinate.

Another hormone, **ethene**, promotes leaf fall (abscission) and fruit ripening. **Abscisic acid** (ABA) works in opposition to gibberellins to stop seed germination.

Revision tip: Concentrate on the units

The amount of a substance is measured in moles, and its concentration tends to be measured in $mol\,dm^{-3}$ (the number of moles in a cubic decimetre of solution). A decimetre is a tenth of a metre (0.1 m), whereas a centimetre is a hundredth of a metre (0.01 m). When dealing with volumes, this means $1\,dm^3 = 1000\,cm^3$. The amount of substance (in moles) can be found by calculating volume × concentration.

Maths skill: Serial dilutions

The effect of plant hormones on growth can be investigated by creating different concentrations of hormone solution. This is achieved using serial dilutions, as illustrated in the following example:

- Solution X has a volume of $10\,cm^3$ and a concentration of $1\,mol\,dm^{-3}$
- $1\,cm^3$ of X is added to $9\,cm^3$ of distilled water. This produces a $0.1\,mol\,dm^{-3}$ solution (Y)
- $1\,cm^3$ of Y is added to $9\,cm^3$ of distilled water. This produces a $0.01\,mol\,dm^{-3}$ solution

Summary questions

1 State three roles of auxins. *(3 marks)*

2 Using your knowledge from Chapter 14, suggest two similarities and two differences between the ways in which plant and animal hormones operate. *(4 marks)*

3 A gibberellin solution with a concentration of $0.5\,mol\,dm^{-3}$ and a volume of $50\,cm^3$ was diluted. $1\,cm^3$ of the solution was added to $9\,cm^3$ of water. From this second solution, $1\,cm^3$ was removed and mixed with $9\,cm^3$ of water. Calculate the concentration and number of moles in the final solution. *(2 marks)*

16.2 Plant responses to abiotic stress
16.3 Plant responses to herbivory

Specification reference: 5.1.5(a)(i) and (b)

Plants experience abiotic stresses. They must respond to environmental fluctuations, such as changes in day length and temperature. They also experience biotic stress – consumption by other organisms is a threat that looms ever-present for many plant species. Plants have evolved responses to cope with the stresses of environmental change and herbivory.

Plant responses

	Factor	Plant response	Mechanism
Abiotic	Day length change	Leaf loss (abscission)	Reduced light exposure → reduced auxin → increased ethene sensitivity → increased cellulase production → cell walls digested at abscission zone.
		Flowering (either when days become shorter or longer)	The relative proportion of two phytochrome photoreceptors (P_{fr} and P_r) dictates when flowering begins. P_{fr} accumulates in the day and is converted back to P_r at night.
	Cold weather	Solutes produced as antifreeze	Reduced temperatures and day length switch certain genes on or off (i.e. epigenetic control of transcription).
	Water shortage	Stomatal closure	The release of ABA causes stomata to close.
Biotic	Herbivory	Physical defences	E.g. the evolution of thorns, barbs, spikes, and inedible tissue.
		Chemical defences	E.g. toxic compounds such as tannins, alkaloids, and terpenoids. Pheromones warn other plants (or regions of the same plant) of an attack.

 Go further: Flowering – the long and the short of it

The balance between P_{fr} and P_r proteins is thought to determine the onset of flowering in plants. P_{fr} is produced during the day. P_r is produced during the night. Some species (long-day plants) flower when day length increases. Other species (short-day plants) flower as day length decreases.

1 Suggest how the relative proportions of P_r and P_{fr} change to initiate flowering in short-day and long-day plants.

Synoptic link

In Topic 9.5, Plant adaptations to water availability, and 12.4, Plant defences against pathogens, you learned about the evolution of other plant adaptations.

Summary questions

1 Describe the defences plants have evolved against herbivory. (*4 marks*)

2 Explain how leaf fall is initiated in some plants as day length decreases. (*4 marks*)

3 Using information from this topic and Topic 12.4, Plant defences against pathogens, suggest how a plant is able to prepare for a pathogenic attack when a neighbouring plant becomes infected. (*2 marks*)

A **tropism** is the growth of a plant in a particular direction. Plants grow towards sources of light, which is known as **phototropism**. Growth in response to gravity is called **geotropism**.

Types of tropism

Tropisms can be **positive** or **negative**. Shoots demonstrate negative geotropism (growing against gravity) and positive phototropism (growing towards light), whereas roots show positive geotropism and negative phototropism.

Question and model answer: Phototropism

Q. Describe the mechanism that controls phototropism in plants, including the role of auxins.

A.

- The shoot **apex** produces **auxin**, which diffuses (or is actively transported) down the shoot.

- Auxin concentration will be **greater on the shaded side** of the shoot/the side receiving less light.

- Auxin causes shaded **cells to elongate**/expand because their **cell walls are loosened**/stretched.

- The cell walls are loosened because **H⁺ ions** are pumped into the walls, **lowering pH**, which increases the activity of **expansins** (proteins that mediate the loosening of cell walls).

- Cells elongate more on the shaded side than the brighter side of the shoot.

- The shoot therefore bends towards the light.

Practical skill: Investigating tropisms

Phototropism can be investigated using the following approaches:

- Vary **light intensity** (e.g. seedlings can be grown in darkness, all-round light, or unilateral light (i.e. light from one side of the plant)).

- Grow seedlings in unilateral light and vary **light wavelengths**, using filters.

- Vary **auxin concentration** (e.g. by removing the shoot apex or introducing additional auxin to one side of the plant).

Geotropism can be demonstrated by rotating a growing plant in a **clinostat**. The shoot and root should grow straight because the gravitational pull is applied evenly to the plant.

Summary questions

1 Describe how geotropism can be demonstrated using a clinostat.
(2 marks)

2 Describe the role of active transport in positive phototropism. *(3 marks)*

3 Explain the importance of positive and negative phototropism to plant survival. *(4 marks)*

16.5 The commercial use of plant hormones

Specification reference: 5.1.5(f)

The production of hormones in plants is often dictated by environmental conditions. For example, ABA production is increased when water is in short supply. However, scientists are now able to harness their knowledge of plant biology and manipulate hormone concentrations for commercial benefit.

How can hormones be used?

Hormone	Commercial uses
Ethene	Speeds up **ripening** Promotes **fruit dropping** (e.g. in walnuts and cherries)
Auxins	Slows down **leaf fall** and **fruit dropping** (although high concentrations actually promote fruit dropping) Encourages **root growth** (in cuttings) Synthetic auxins can act as **weedkillers**
Cytokinins	**Prevents senescence** (ageing)
Gibberellins	**Delays senescence** (in citrus fruit) Speeds up barley **seed germination** during alcohol brewing

Synoptic link

In later topics (e.g. Topic 22.1, Natural cloning in plants, and Topic 22.2, Artificial cloning in plants) you will examine other uses of plant hormones.

Summary questions

1 State three commercial uses of auxin. *(3 marks)*

2 Solutions of 2-chloroethylphosphonic acid are often applied to plants instead of ethene, which is a gas. Suggest why. *(2 marks)*

3 What can you conclude from Figure 1 about the effect of ethene on fruit ripening? Explain your answer. *(2 marks)*

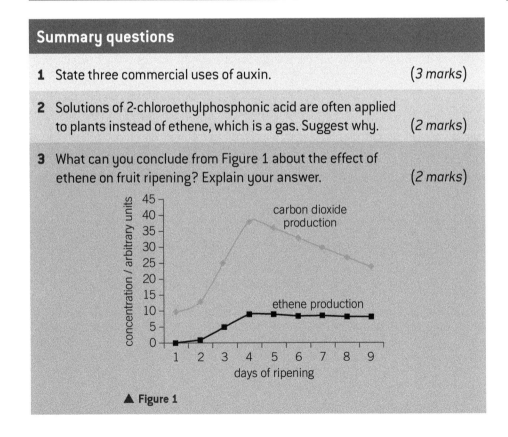

▲ Figure 1

1 Which of the following scenarios is/are likely to result in flowering in short-day plants? *(1 mark)*

   1 An increase in night length beyond a critical period.

   2 An increase in day length beyond a critical period.

   3 An increase in P_{fr} concentration within the plant.

   A 1, 2, and 3 are correct

   B Only 1 and 2 are correct

   C Only 2 and 3 are correct

   D Only 1 is correct

2 Which of the following statements is/are true of gibberellins? *(1 mark)*

   1 They promote seed germination.

   2 They promote stem elongation.

   3 Their action is inhibited by ABA.

   A 1, 2, and 3 are correct

   B Only 1 and 2 are correct

   C Only 2 and 3 are correct

   D Only 1 is correct

3 The graph below shows the effect of gibberellin exposure on amylase production in isolated tissue from barley seeds.

   Describe and explain the data shown in the graph. *(5 marks)*

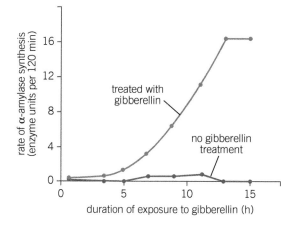

4 Suggest the significance of the data in the graph below to scientists carrying out tissue culture. *(2 marks)*

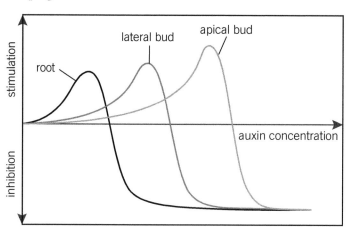

17.1 Energy cycles
Specification reference: 5.2.1(a) and 5.2.2(a)

Energy is required for many biological processes, including active transport, movement, and anabolic reactions. Energy enters ecosystems through photosynthesis, in which plants trap solar energy as chemical energy. Cellular respiration converts the chemical energy to a useable form: ATP. Here you will examine the relationship between photosynthesis and respiration.

A comparison of respiration and photosynthesis

Although the two processes involve many different reactions, the overall equations for photosynthesis and respiration mirror each other. Photosynthesis uses light energy to convert inorganic molecules (carbon dioxide and water) to organic molecules (e.g. glucose) and oxygen. Respiration converts glucose back to carbon dioxide and water, which releases the energy that is used to generate ATP.

	Photosynthesis	Respiration
Purpose	Conversion of light energy to chemical energy in organic molecules $6CO_2 + 6H_2O \rightarrow C_6H_{12}O_6 + 6O_2$	Conversion of chemical energy in organic molecules to chemical energy in ATP $C_6H_{12}O_6 + 6O_2 \rightarrow 6CO_2 + 6H_2O$
Reactants	Carbon dioxide and water	Glucose and oxygen
Products	Glucose and oxygen	Carbon dioxide and water
Type of reaction	Endothermic (overall, energy is taken in)	Exothermic (overall, energy is released)
ATP production	Produced in the light-dependent stage and used in the light-independent stage	An end product
Use of coenzymes	NADP carries H atoms between the two stages of photosynthesis	NAD and FAD carry H atoms to the electron transport chain

Key term

Photosynthesis: The process by which plants harness light to produce complex organic molecules from carbon dioxide and water.

Key term

Respiration: The breakdown of complex organic molecules to produce ATP.

Revision tip: Energy is transferred not lost

The amount of energy in the universe has remained the same since the Big Bang. This energy can be released and absorbed, or converted to different forms. Avoid writing that energy is produced, made, or lost.

Synoptic link

You will learn about how energy is transferred through ecosystems in Topic 23.2, Biomass transfer through an ecosystem.

Summary questions

1 Complete the following passage by choosing the most appropriate words to place in the gaps.
Photosynthesis is .... (i.e. energy is taken in, overall).
It converts light energy to chemical energy, which is required by organisms for processes such as .... transport and anabolic reactions. One product of photosynthesis is glucose, which is broken down during cellular respiration to produce ....; this molecule is the currency of chemical energy used by all cells. (3 marks)

2 State three similarities between the processes of photosynthesis and respiration. (3 marks)

3 Describe the energy conversions and transfers that occur between light being absorbed by chlorophyll in plant leaves and heat being radiated from organisms. (3 marks)

17.2 ATP synthesis

Specification reference: 5.2.2(h)

Synoptic link

Topic 3.11 , ATP, outlined the structure of ATP.

ATP is the currency of chemical energy used by all species. Cellular respiration generates ATP as an end product. However, ATP is also produced by the light-dependent stage of photosynthesis. In both cases, ATP production relies on electron transport chains, chemiosmosis, and ATP synthase.

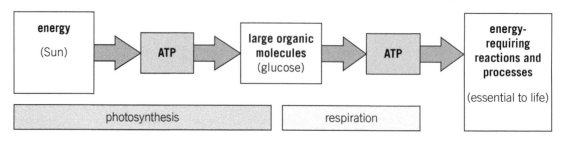

▲ **Figure 1** *An overview of ATP's role in photosynthesis and respiration*

Key term

Chemiosmosis: The movement of H⁺ ions down an electrochemical gradient, which drives the production of ATP in both respiration and photosynthesis.

Summary questions

1 State two differences between chemiosmosis in respiration and photosynthesis. (*2 marks*)

2 Describe how a proton gradient is established across a thylakoid membrane during photosynthesis. (*3 marks*)

3 Suggest how the following research findings provide evidence for the chemiosmotic theory.
   a The pH of the mitochondrial intermembrane space is lower than that of the matrix. (*2 marks*)
   b The potential difference across the inner membrane is −200 mV. The matrix is more negative than the intermembrane space. (*1 mark*)

Chemiosmosis

In both photosynthesis and respiration, chemiosmosis involves the following steps:

1 Electrons are raised to a higher energy level (i.e. **excited electrons**)

2 The high-energy electrons pass along an **electron transport chain** (i.e. a series of electron carriers)

3 Energy is released as the electrons are passed to lower energy levels

4 The energy is used to pump H⁺ ions (protons) across a membrane

5 A **proton gradient** is established across the membrane

6 Protons move down the concentration gradient through channel proteins linked to **ATP synthase**

7 The flow of protons provides kinetic energy to enable **ATP synthesis** by ATP synthase.

Comparison of chemiosmosis in respiration and photosynthesis

The principles of chemiosmosis are the same in both respiration and photosynthesis, but several differences exist.

▼ **Table 1** *A comparison of chemiosmosis in photosynthesis and respiration*

	Photosynthesis	Respiration
Where do the high energy electrons come from?	Light is absorbed by chlorophyll	Electrons are released from chemical bonds in glucose
Where is the location of the electron transport chain?	Thylakoid membranes (in chloroplasts)	Inner mitochondrial membranes
What is the name given to ATP production?	Photophosphorylation	Oxidative phosphorylation

17.3 Photosynthesis

Specification reference: 5.2.1(b), (c)i-ii, (d), (e), and (f)

Photosynthesis comprises two stages, both of which occur in chloroplasts. The light-dependent stage relies on a range of pigments to absorb sunlight and convert it to chemical energy. The light-independent stage fixes carbon dioxide and transfers the chemical energy into organic molecules such as glucose.

Chloroplast structure

Chloroplasts consist of two distinct regions:

- **Grana** (singular: granum): flattened membrane compartments (thylakoids), which are the sites of the light-dependent stage of photosynthesis.

- **Stroma**: A fluid-filled matrix, which is the site of the light-independent stage of photosynthesis.

Synoptic link

You learned about the structure of chloroplasts in Topic 2.5, The ultrastructure of plant cells.

▼ **Table 1** *The adaptations of chloroplasts*

Adaptation	Purpose
Thylakoid membranes are stacked	Large surface area over which light-dependent reactions can occur
Photosynthetic pigments are organised into photosystems	The efficiency of light absorption is maximised
Grana are surrounded by the stroma	Products of the light-dependent reactions (reduced NADP and ATP) can pass directly to the enzymes catalysing the light-independent reactions
Chloroplasts contain their own DNA and ribosomes	Photosynthetic proteins can be produced inside chloroplasts rather than being imported
The inner chloroplast membrane is embedded with transport proteins and is less permeable than the outer membrane	Control over which substances that can enter the stroma from the cell cytoplasm

Photosynthetic pigments

- **Photosystems** are light-harvesting complexes of pigments found in thylakoid membranes.

- **Accessory pigments** (e.g. chlorophyll *b*, carotenoids, and xanthophylls) absorb photons of light and funnel this energy to a **reaction centre** at the heart of the photosystem.

- **Chlorophyll *a*** is located in the reaction centre.

- **Electrons** from chlorophyll *a* are excited and passed to electron acceptors at the beginning of the electron transport chain.

Practical skill: Thin layer chromatography

Photosynthetic pigments can be separated and identified using a technique called **chromatography**.

How are pigments separated? The pigment molecules interact with the stationary phase to different extents and therefore move at different rates.

What is the stationary phase? Either silica gel on glass, or paper.

What is the mobile phase? A solvent in which the pigment molecules dissolve.

Pigments can be identified and compared by calculating R_f values:

$$R_f = \frac{\text{distance travelled by pigment}}{\text{distance travelled by solvent}}$$

▼ **Table 2** *An example of R_f calculations*

Pigment	Distance travelled (cm)	Solvent distance (cm)	R_f
carotene	4.75	5.00	0.95
chlorophyll a	3.25	5.00	0.65
chlorophyll b	2.25	5.00	0.45

The light-dependent stage

The purpose of the light-dependent stage of photosynthesis is to generate two products, which then feed into the **Calvin cycle**.

- **Light energy** is absorbed and enables **ATP production**.
- **Hydrogen from water** is used to **reduce NADP**.

Non-cyclic phosphorylation in the light-dependent stage involves the following steps:

1 Electrons from the reaction centre in photosystem II are excited

2 The excited electrons are passed along an electron transport chain (and ATP is produced)

3 The electrons from photolysis (the splitting of water using light energy) replace those lost from photosystem II

4 Electrons from the reaction centre in photosystem I are excited

5 More ATP is produced via a second electron transport chain

6 The electrons that left photosystem II replace those lost from photosystem I

7 The electrons from photosystem I and H+ ions from the photolysis of water combine to reduce NADP.

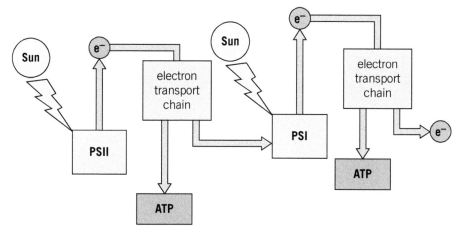

▲ **Figure 1** *Non-cyclic phosphorylation*

The light-independent stage (the Calvin cycle)

Hydrogen (from reduced NADP) and carbon dioxide are used to synthesise glucose and other organic molecules. The energy required for these reactions is supplied by ATP.

The Calvin cycle involves the following steps:

1 The enzyme RuBisCo catalyses the reaction between RuBP (a 5-carbon sugar) and CO_2 to produce two molecules of three-carbon GP

2 ATP and NADPH are used to reduce GP to another three-carbon molecule, TP

3 The majority of TP is used to regenerate RuBP, which continues the cycle. More ATP is required for this reaction.

Uses of triose phosphate

Other than being used to regenerate RuBP, triose phosphate (TP) can be converted into a wide range of products:

- Six-carbon sugars (e.g. glucose and fructose)
- Fructose and glucose can react to produce the disaccharide sucrose
- Glucose molecules can react together to form polysaccharides such as amylose, amylopectin, and cellulose
- A single TP molecule can be converted to glycerol
- TP can act as the starting point for the synthesis of amino acids.

Revision tip: What's in a name?

You might see the molecules of the Calvin cycle being referred to by a variety of names, depending on the book or article you are reading. Glycerate-3-phosphate (GP) is sometimes called PGA or 3PG. Triose phosphate (TP) is sometimes called G3P, GALP, or PGAL.

➕ Go further: A visit to the GP

Figure 2 shows some of the molecules that can be formed from GP. We can see that only small changes occur during these reactions. When GP is converted to the intermediate 1,3-diphosphoglycerate, the OH (hydroxyl) group is replaced by a phosphate group. When 1,3-diphosphoglycerate is converted to TP, this phosphate is substituted for a hydrogen atom. Figure 2 also shows the conversion of GP into another important biological molecule, serine.

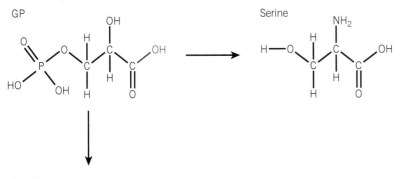

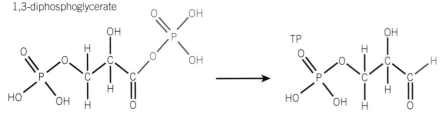

▲ **Figure 2** *An outline of molecules that can be produced from GP*

1. Which molecules are required for the conversion of (a) GP to 1,3-diphosphoglycerate (b) 1,3-diphosphoglycerate to TP?

2. Serine is an amino acid. Draw the R group.

3. Suggest a mineral ion that would be required during the conversion of GP to serine.

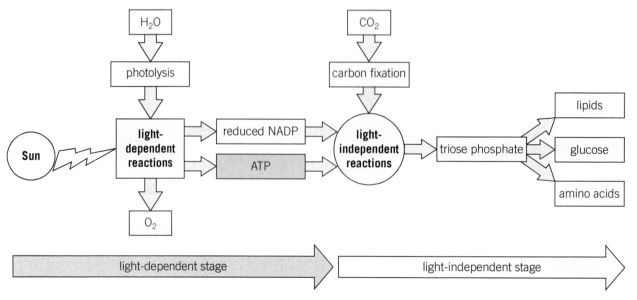

▲ **Figure 3** *An overview of photosynthesis*

Summary questions

1 Name two useful molecules that can be produced from TP and describe one way in which a plant can use each molecule. (*4 marks*)
2 Figure 4 shows a chromatogram of two chloroplast pigments. Calculate R_f values for pigments A and B. (*2 marks*)

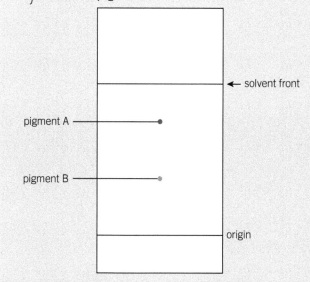

▲ **Figure 4** *Chromatogram of chloroplast pigments*

3 Explain why the Calvin cycle will stop after a plant has been placed in the dark for a period of time. (*2 marks*)
4 One glucose molecule can be produced for every six turns of the Calvin cycle. The mass of one molecule of RuBP is approximately 1.7 times greater than the mass of one molecule of glucose. Estimate the mass of RuBP required to produce 10 kg of glucose. Give your answer to three significant figures. (*2 marks*)

5 Suggest three reasons why plants cannot use the ATP produced in the light-dependent stage of photosynthesis as their only source of ATP. (*3 marks*)

17.4 Factors affecting photosynthesis

Specification reference: 5.2.1(g)i-ii

Several environmental factors influence the rate of photosynthesis. In this topic you will consider some of these limiting factors and the methods that are available to study them.

Limiting factors in photosynthesis

Limiting factor	How does it affect photosynthetic rate?
Light intensity	Light intensity determines the rate at which ATP and reduced NADP are produced in the light-dependent reactions.
Carbon dioxide concentration	Low CO_2 concentrations slow the rate of GP formation.
Temperature	Low temperatures limit the kinetic energy of molecules involved in photosynthetic reactions. High temperatures can cause enzymes such as RuBisCo to denature.

Practical skill: Investigating factors that affect photosynthetic rate

Photosynthesis can be measured using a variety of approaches. **Sensors** and **data loggers** can be used to monitor the effect of various factors on photosynthetic rate.

Several factors can act as independent variables. These include temperature, light intensity, and CO_2 concentration. Remember, in a well-designed experiment only the independent variable will change. All other factors must be controlled. You will therefore be testing the effect of a single factor (the independent variable) on the rate of photosynthesis (the dependent variable).

Summary questions

1   Figure 1 shows the effect of carbon dioxide concentration on RuBP and GP concentrations. Left of the central line = 1.0% carbon dioxide. Right of the central line = 0.003% carbon dioxide. Explain the graph.   (*2 marks*)

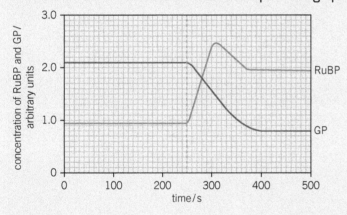

▲ Figure 1

2   Suggest why temperature changes have a greater impact on the rate of the light-independent reactions than the rate of the light-dependent stage of photosynthesis.   (*2 marks*)

3   Figure 2 shows the effect of light intensity on GP, TP, and RuBP concentrations. Explain the graph.   (*3 marks*)

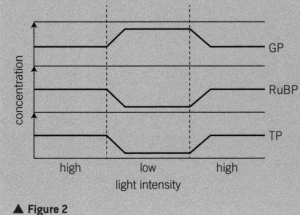

▲ Figure 2

1   A pigment has an R$_f$ value of 0.60. A student uses paper chromatography. The pigment travels 5.33 cm. How far (to two significant figures) does the solvent in which the pigment is dissolved travel?   (*1 mark*)

   A 3.1 cm   **C** 8.8 cm

   B 3.2 cm   **D** 8.9 cm

2   One glucose molecule is produced for every six turns of the Calvin cycle. The mass of one molecule of RuBP is approximately 1.7 times greater than the mass of one molecule of glucose. What is the mass of RuBP, to 2 significant figures, required to produce 5.75 kg of glucose?   (*1 mark*)

   A 5800 g   **C** 58 000 g

   B 5900 g   **D** 59 000 g

3   A potometer was used to investigate the effect of the temperature on photosynthetic rate. At 20°C, an oxygen bubble 2.5 cm long was collected over a one-minute period. The radius of the potometer was 0.07 cm. What was the rate of photosynthesis to 2 significant figures?   (*1 mark*)

   A 0.038 cm³ min⁻¹   **C** 0.54 cm³ min⁻¹

   B 0.040 cm³ min⁻¹   **D** 0.55 cm³ min⁻¹

4   Discuss the methods that can be employed on farms to increase primary production in crop plants.   (*6 marks*)

5   Describe and explain the shape of the graph below.   (*3 marks*)

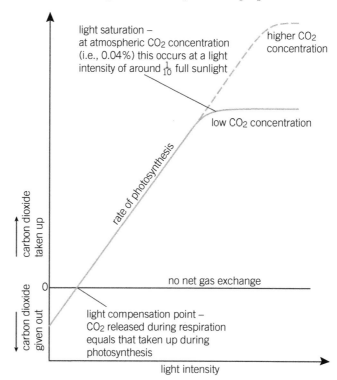

18.1 Glycolysis

Specification reference: 5.2.2(c)

You have learned how energy is trapped in organic molecules by plants. Organic molecules are metabolised to release useable energy in a process called cellular respiration. Here you will examine the first stage of respiration: glycolysis.

Revision tip: Step by step ...

Glycolysis involves ten steps, but you are required to know only that glucose is converted to hexose bisphosphate, which splits into two TP molecules, which are oxidised to two pyruvate molecules.

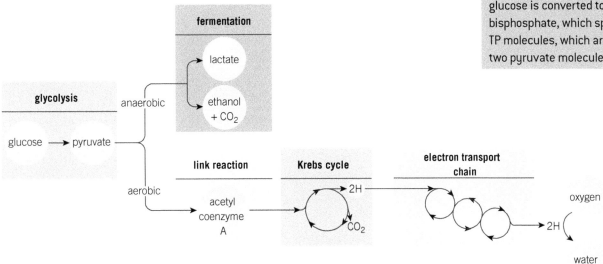

▲ **Figure 1** *An overview of respiration*

Glycolysis

Glycolysis occurs in the **cytoplasm** of cells and comprises the following steps:

1 Glucose is **phosphorylated** to hexose bisphosphate by two **ATP molecules**. This prevents glucose molecules leaving the cell and destabilises them, which makes them easier to break down.

2 Hexose bisphosphate (a six-carbon sugar) **splits** into two molecules of **triose phosphate** (TP – which are three-carbon molecules).

3 TP is **phosphorylated** (by free inorganic phosphate rather than ATP).

4 The two TP molecules lose H atoms (**dehydrogenation**, forming reduced NAD) and phosphate (enabling **ATP formation**) to produce two molecules of **pyruvate**.

▼ **Table 1** *A summary of reactants and products in glycolysis (per glucose molecule)*

Used	Produced
Glucose **2 ATP**	**2 pyruvate molecules** (*which are either used in the link reaction in mitochondria or fermented to enable anaerobic respiration*) **4 ATP** (*through substrate-level phosphorylation*) **2 reduced NAD** (*which are used in the electron transport chain*)

Synoptic link

TP is a molecule produced in the Calvin cycle, which you encountered in Topic 17.3, Photosynthesis. You will learn about the importance of glycolysis for anaerobic respiration in Topic 18.5, Anaerobic respiration.

Summary questions

1 Explain why one glucose molecule yields two molecules of ATP from glycolysis. *(3 marks)*

2 Using information in Topic 18.4, Oxidative phosphorylation, estimate the maximum number of ATP molecules that can be generated from one glucose molecule passing through glycolysis. Explain the source of the ATP molecules. *(4 marks)*

3 Describe the role of phosphorylation in glycolysis. *(4 marks)*

18.2 Linking glycolysis and the Krebs cycle

Specification reference: 5.2.2(b) and (d)

The three-carbon pyruvate molecules produced in glycolysis need to be converted to two-carbon molecules before they can enter the Krebs cycle. This conversion is known as the link reaction.

The link reaction

The link reaction occurs in the **matrix** of mitochondria. This requires pyruvate to move from the cytoplasm into mitochondria through active transport. Three things happen in the link reaction:

- Pyruvate loses hydrogen (*dehydrogenation*), which produces reduced NAD.
- Pyruvate loses carbon (*decarboxylation*), which produces CO_2 and a two-carbon group called acetyl.
- The acetyl group binds to coenzyme A, which produces acetyl coenzyme A (*acetyl coA*).

The role of coenzyme A is to deliver the acetyl group to the next stage of respiration, the **Krebs cycle**.

▼ **Table 1** *A summary of reactants and products in the link reaction (per glucose molecule)*

Used	Produced
2 pyruvate molecules	2 CO_2
	2 reduced NAD (which are used in the electron transport chain)
	2 acetyl coA

Revision tip: Acetyl vs. acetate

The group that binds to coenzyme A is acetyl (CH_3CO), not acetate (CH_3COO^-).

Synoptic link

You learned about coenzymes in Topic 4.4, Cofactors, coenzymes, and prosthetic groups.

Revision tip: Where there's oxidation, there's reduction

The dehydrogenation of pyruvate is an example of a redox reaction. Pyruvate loses hydrogen and is oxidised, and NAD gains hydrogen and is reduced.

Pyruvate + coenzyme A → acetyl coenzyme A

NAD → reduced NAD

CO_2

Mitochondrion structure

outer mitochondrial membrane separates the contents of the mitochondrion from the rest of the cell, creating a cellular compartment with ideal conditions for aerobic respiration

inner mitochondrial membrane contains electron transport chains and ATP synthase

cristae are projections of the inner membrane which increase the surface area available for oxidative phosphorylation

matrix contains enzymes for the Krebs cycle and the link reaction, also contains mitochondrial DNA

intermembrane space Proteins are pumped into this space by the electron transport chain. The space is small so the concentration builds up quickly

▲ **Figure 1** *The structure of a mitochondrion*

Summary questions

1 Describe what happens to pyruvate after it has been produced in glycolysis. (4 marks)

2 Explain what happens to the carbon dioxide produced by the link reaction in plant species. (2 marks)

3 Suggest why pyruvate is transported into mitochondria after it is produced in glycolysis. (2 marks)

18.3 The Krebs cycle

Specification reference: 5.2.2(e) and (f)

Following the link reaction, acetyl CoA enters the Krebs cycle. This cycle comprises a series of reactions that generate ATP and reduced coenzymes (reduced FAD and NAD). The reduced coenzymes enable even more ATP to be produced by oxidative phosphorylation. Carbon dioxide is released as a by-product of the cycle.

The Krebs (citric acid) cycle

As with the link reaction, the reactions of the Krebs cycle occur in the mitochondrial matrix. The key reactions in the cycle are:

1 Acetyl CoA delivers an **acetyl group**.

2 Acetyl (two-carbon) reacts with **oxaloacetate** (four-carbon) to produce **citrate** (six-carbon).

3 Citrate loses H (*dehydrogenation*, forming reduced NAD) and C (*decarboxylation*, forming CO_2). This results in a five-carbon compound being produced.

4 The five-carbon compound is decarboxylated and dehydrogenated further, regenerating oxaloacetate, which is free to react with another acetyl CoA.

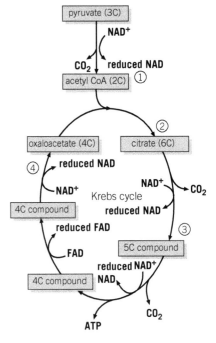

▲ Figure 1 *The Krebs cycle*

▼ **Table 1** *A summary of reactants and products in the Krebs cycle (per glucose molecule)*

Used	Produced
2 Acetyl-CoA	4 CO_2 6 reduced NAD *(which are used in the electron transport chain)* 2 reduced FAD *(which are used in the electron transport chain)* 2 ATP *(through substrate-level phosphorylation)*

Revision tip: Just a FAD ... or is it a NAD?

NAD and FAD are two more examples of coenzymes. They both carry protons and electrons from earlier stages of respiration to the electron transport chain. This enables ATP generation. However, as you will see in the next topic, more ATP can be produced from a NAD molecule than a FAD molecule.

Revision tip: Remember these two

Citrate and oxaloacetate are the only Krebs cycle molecules you need to learn. Names of the other intermediates are not required.

Summary questions

1 How many of the original six carbon atoms in a glucose molecule are released as carbon dioxide during the Krebs cycle? *(1 mark)*

2 Using information in Topic 18.4, Oxidative phosphorylation, estimate the maximum number of ATP molecules that can be generated from one turn of the Krebs cycle. Explain the source of the ATP molecules. *(4 marks)*

3 Suggest why oxaloacetate is present in cells at very low concentrations. *(2 marks)*

18.4 Oxidative phosphorylation

Most of the ATP generated in respiration is produced via oxidative phosphorylation. The reduced coenzymes ($FADH_2$ and NADH) produced during earlier stages of respiration donate H^+ ions and electrons to an electron transport chain. ATP is synthesised by chemiosmosis.

The electron transport chain

Electron transport chains (ETCs) are located on inner mitochondrial membranes. The chains use energy from electrons to pump H^+ ions into the intermembrane space. A proton gradient is established, which enables chemiosmosis to occur through ATP synthase. The sequence of events is as follows:

1 High energy **electrons** are passed from NADH and $FADH_2$ to **electron carriers** in the ETC.

2 Electrons are passed between carriers in a series of **redox reactions**.

3 Energy is released from each reaction.

4 The energy is used to **pump H^+ ions** from the matrix into the intermembrane space.

5 H^+ ions diffuse through ATP synthase (**chemiosmosis**) back into the matrix, which produces ATP.

6 H^+ ions and electrons react with O_2 to produce water.

Revision tip: Just a number...

You are not required to learn how many ATP molecules are generated in oxidative phosphorylation. However, you might be provided with relevant data and asked to interpret the information.

▼ **Table 1** *A summary of reactants and products in oxidative phosphorylation (per glucose molecule)*

Used	Produced
10 reduced NAD (NADH) 2 reduced FAD ($FADH_2$) 6 O_2	26–34 ATP (estimates vary – see Go further for more details) 6 H_2O

Synoptic link

You examined the principles of chemiosmosis in Topic 17.2, ATP synthesis.

Summary questions

1 Explain how the structure of mitochondrial cristae maintains a high rate of oxidative phosphorylation. *(2 marks)*

2 Explain why oxygen is called the final electron acceptor in respiration. *(3 marks)*

3 The theoretical maximum yield of ATP from aerobic respiration may not always be achieved. Suggest why. *(2 marks)*

Go further: How many ATP molecules are produced per glucose molecule?

Scientists have calculated different estimates of the number of ATP molecules produced per glucose molecule. Some suggest 38 is the theoretical maximum yield. However, the current prevailing opinion is that 30 or 32 ATP molecules are produced per glucose. The crux of the calculation is working out how many ATPs are generated for each NADH and $FADH_2$. Most scientists think that NADH produces 2.5 ATPs, whereas $FADH_2$, because it enters the ETC at a later stage, produces only 1.5. The following table summarises current understanding:

Stage of respiration	Source of ATP	Number of ATP molecules
Glycolysis	Substrate-level phosphorylation	2
	2 NADH (oxidative phosphorylation)	3–5
Link reaction	2 NADH (oxidative phosphorylation)	5
Krebs cycle	Substrate-level phosphorylation	2
	2 NADH (oxidative phosphorylation)	15
	2 $FADH_2$ (oxidative phosphorylation)	3
Total ATP yield (per glucose)		30–32

1 Suggest why only 1.5 ATP molecules might be produced from a molecule of reduced NAD generated during glycolysis.

2 Use the following information to determine where reduced NAD and reduced FAD enter the electron transport chain:

Four protein carriers are present. Carrier 1 and 3 each pump 4 H^+ ions per coenzyme. Carrier 2 pumps none, and carrier 4 pumps 2 H^+ ions into the intermembrane space. Scientists think that 4 H^+ ions are required to pass through ATP synthase to synthesise one molecule of ATP.

18.5 Anaerobic respiration

Specification reference: 5.2.2(i)

Oxygen, as you learned in the previous topic, is the electron acceptor at the end of respiration. ATP cannot be produced via oxidative phosphorylation without oxygen. However, organisms can still generate some ATP even when oxygen is unavailable.

Anaerobic respiration

Anaerobic respiration generates small amounts of ATP through **substrate-level phosphorylation** in **glycolysis**. The link reaction, the Krebs cycle, and the electron transport chain all stop working in the absence of oxygen.

Fermentation

The function of fermentation is to oxidise NADH, thereby regenerating NAD. This enables glycolysis to continue. Glycolysis would stop without fermentation because supplies of NAD would run out, and TP would no longer be converted to pyruvate.

Fermentation in **mammals** involves reduced NAD donating H atoms to pyruvate, producing lactate:

Pyruvate + reduced NAD → lactate + NAD

Fermentation in **yeast** and **plants** involves pyruvate breaking down to ethanal and CO_2. Reduced NAD then donates H atoms to ethanal, producing ethanol:

Pyruvate → ethanal + CO_2   *Ethanal + reduced NAD → ethanol + NAD*

Practical skill: Measuring anaerobic respiration rates

The rate of anaerobic respiration in yeast can be investigated using the following method:

- Yeast cultures are placed in sealed flasks (*to ensure anaerobic conditions*).

- The **dependent variable** is the rate of carbon dioxide production (*which can be measured by monitoring the displacement of coloured liquid in a capillary tube; this is similar to the use of a potometer, which you learned about in Topic 9.3, Transpiration*).

- When testing the effect of an **independent variable** (e.g. temperature) on respiration rate, only that variable should be changed. All other factors (e.g. pH, glucose concentration) are **controlled.**

Summary questions

1 Explain why mammals use anaerobic respiration only for short periods of time. *(2 marks)*

2 The hydrolysis of one mole of glucose releases 2880 kJ of energy. The hydrolysis of one mole of ATP releases 30.6 kJ of energy. For every mole of glucose, aerobic respiration produces 32 moles of ATP, whereas anaerobic respiration produces 2 moles of ATP. Calculate the percentage efficiencies of the two forms of respiration. *(2 marks)*

3 Suggest why the absence of oxygen stops
   a chemiosmosis and the electron transport chain *(2 marks)*
   b the Krebs cycle. *(2 marks)*

18.6 Respiratory substrates

Molecules other than glucose can be used as respiratory substrates. Amino acids, glycerol, and fatty acids can all be metabolised and enter respiration at different points.

Respiratory substrates

The following table outlines how alternative respiratory substrates are used.

Respiratory substrate		Which stage of respiration does it enter?	Which molecule is it used to form?
Amino acids		The link reaction or the Krebs cycle	This depends on the amino acid e.g. Glycine $\rightarrow$ pyruvate Isoleucine $\rightarrow$ acetyl CoA Aspartate $\rightarrow$ oxaloacetate
Triglycerides	Glycerol	Glycolysis	Triose phosphate
	Fatty acids	Krebs cycle	Acetyl CoA
Lactate		Link reaction	Pyruvate

 Worked example: Respiratory quotient

$$\text{Respiratory quotient (RQ)} = \frac{\text{Carbon dioxide (produced)}}{\text{Oxygen (consumed)}}$$

How do you work out RQ for a particular respiratory substrate?

Let us examine the respiration of stearic acid $(C_{17}H_{35}COOH)$ as an example.

1 **Write an equation.** $C_{17}H_{35}COOH + (?)O_2 \rightarrow (?)CO_2 + (?)H_2O$

2 The number of CO_2 = number of C atoms in the substrate.
$C_{17}H_{35}COOH + (?)O_2 \rightarrow \mathbf{18}CO_2 + (?)H_2O$

3 The number of H_2O = half the H atoms in the substrate.
$C_{17}H_{35}COOH + (?)O_2 \rightarrow 18CO_2 + \mathbf{18}H_2O$

4 Balance the number of oxygen atoms on each side.
$C_{17}H_{35}COOH + \mathbf{26}O_2 \rightarrow 18CO_2 + 18H_2O$

5 Calculate RQ: 18/26 = 0.69.

Revision tip: What does RQ tell us?

RQ of more than 1 = some carbon dioxide is produced from anaerobic respiration

RQ of 1 or less = aerobic respiration

An RQ of 1 indicates carbohydrates are being respired, 0.8–0.9 indicates proteins, and approximately 0.7 indicates lipids.

Summary questions

1 What must be removed from all amino acids before they can be metabolised as respiratory substrates? State the name of this process. *(2 marks)*

2 Calculate the RQ of palmitic acid $(C_{15}H_{31}COOH)$. Give your answer to three significant figures. *(3 marks)*

3 Some respiratory molecules are easier than others to form from an amino acid. Suggest the likely respiratory molecules formed from the following amino acids. Explain your answers.
   a asparagine $(C_4H_8N_2O_3)$
   b alanine $(C_3H_7NO_2)$. *(4 marks)*

1   Which of the following statements is/are true of the conversion of
    pyruvate to acetyl coenzyme A in the link reaction?   (*1 mark*)

    1   Pyruvate is reduced.

    2   Pyruvate is decarboxylated.

    3   Pyruvate is dehydrogenated.

    A   1, 2, and 3 are correct

    B   Only 1 and 2 are correct

    C   Only 2 and 3 are correct

    D   Only 1 is correct

2   Which of the following statements is/are true of the electron
    transport chain in mitochondria?   (*1 mark*)

    1   The first electron carrier is reduced by NADH.

    2   Oxygen acts as the terminal electron acceptor.

    3   Energy is used to pump protons out of the intermembrane space.

    A   1, 2, and 3 are correct

    B   Only 1 and 2 are correct

    C   Only 2 and 3 are correct

    D   Only 1 is correct

3   Which molecule acts as the hydrogen acceptor during alcoholic
    fermentation?   (*1 mark*)

    A   Glucose   C   Ethanol

    B   Pyruvate   D   Ethanal

4   What is the respiratory quotient of palmitic acid ($CH_3(CH_2)_{14}COOH$)
    to two significant figures?   (*1 mark*)

    A   0.69

    B   0.70

    C   0.71

    D   0.72

5   Describe how the respiratory quotient of a plant can be
    determined using a respirometer.   (*6 marks*)

6   The graph below illustrates the release of energy during respiration.
    Explain the difference between respiration and combustion in terms of the
    release of chemical energy and its significance.   (*3 marks*)

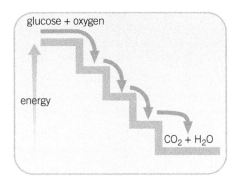

Changes to the base sequences in DNA are known as **mutations**. These alterations vary in their severity. They can be harmful, beneficial, or have no effect.

Types of mutation

You can consider mutations in terms of how the DNA changes or the effect of the mutation.

Change to DNA	Description	Examples of genetic diseases
Substitution	One nucleotide is exchanged for another.	Sickle cell anaemia
Insertion	An extra nucleotide (or more than one) is placed in the DNA sequence.	Huntington's disease
Deletion	A nucleotide (or more than one) is removed from the DNA sequence.	Cystic fibrosis

Revision tip: Frameshifts

Substitution mutations change one codon and cause a maximum of one amino acid to be altered in a protein's primary structure. Insertion and deletion mutations can change many codons along the DNA sequence; this is called a **frameshift** mutation. These mutations tend to be much more harmful than substitution mutations.

Revision tip: Remaining neutral?

Some mutations change the primary structure of a protein but have no significant effect on its function (i.e. they have a neutral effect).

Type of mutation	Description	Effects
Silent	The mutation could: • be in a **non-coding region** • produce a new codon that codes for the same amino acid (because of the **degenerate code**).	No effect (neutral)
Nonsense	A codon is changed to a **stop codon**.	A shorter polypeptide Normally harmful
Missense	At least one new amino acid is introduced into a protein's primary structure.	Harmful usually, but sometimes beneficial or neutral

Revision tip: Evolution requires mutation

You learned in Topic 10.8, Changing population characteristics, that natural selection requires variation within a population. This variation can be attributed to DNA mutations, which produce variants of genes (i.e. alleles). Occasionally a slight change in a gene can be beneficial. It may improve the function of the polypeptide for which it codes. These slight alterations to DNA sequences tend to be substitution mutations rather than insertions or deletions.

Go further: The FTO 'hunger' gene: an example of the subtle effects of gene mutation?

The FTO gene codes for a protein that regulates appetite and feeding behaviour. Mutations have produced several known variants of the gene. One of these alleles has been linked to an increased risk of obesity.

In a study, people who are homozygous for the risk allele were found to be 3 kg heavier on average and were 70% more likely to have obesity than people without a copy of this allele. A later study found that people with the risk allele have higher levels of the 'hunger hormone' ghrelin in their blood, which means they begin to feel hungry sooner than other people. Ghrelin is coded for by a separate gene.

1 Suggest the type of mutation responsible for the FTO gene variants. Explain your reasoning.

2 Suggest how the protein produced by the FTO gene might control feeding behaviour.

Synoptic link

You looked at the structure of DNA and the degeneracy of the genetic code in Chapter 3, Biological molecules.

Revision tip: Mutagens

DNA mutations can occur spontaneously, but the rate of mutation is increased by mutagens, which can be biological (e.g. viruses), chemical , or physical (e.g. radiation).

Summary questions

1 Explain why an insertion mutation of three nucleotides does not cause a frameshift.   *(3 marks)*

2 Look at the base sequence below. How many triplet codes in this sequence would be changed by
   a all guanine bases experiencing substitution mutations?
   b a substitution mutation of A?
   c a deletion of A?   *(3 marks)*
   G C A C T T G G G C C T C

3 Aromatic rice is a popular cooking ingredient. A 'fragrance' gene has been sequenced in rice plants and found to have two variants: one that results in an aroma and one that does not. Sections of the two gene variants are shown below. The fragrant variant has four separate mutations that make it different to the non-fragrant variant. Identify and describe the four mutations.   *(4 marks)*
   Non-fragrant variant: A A A C T G G T A A A A A G A T T A T G G C T T C A G C T G
   Fragrant variant: A A A C T G G T A T A T A T T T C A G C T G

19.2 Control of gene expression
Specification reference: 6.1.1(b)

A gene is expressed when its sequence of codons is transcribed and translated into a polypeptide. Gene expression can be controlled in several ways, including the prevention of transcription and the post-transcriptional modification of mRNA. Even after translation, a polypeptide can be modified to change its function.

How is gene expression controlled?

The table below outlines how gene expression can be controlled at different stages of polypeptide production.

Stage	Control mechanism	Description
Transcription	Chromatin structural changes	**Heterochromatin** forms during cell division. DNA is wound tightly around histones. No transcription occurs. **Euchromatin** forms during interphase. DNA is wound loosely around histones. Transcription can occur.
	Transcription factors	Molecules that bind to DNA to either promote or prevent transcription.
	Epigenetics	**Acetylation** of histones increases transcription rates. **Methylation** of DNA prevents transcription.
	Operons (e.g. the *lac operon* in prokaryotes, see Figure 1)	Genes are switched *off* when a **repressor** binds to an **operator** region, which blocks a **promoter** region. This prevents **RNA polymerase** binding and stops the transcription of **structural genes**. Genes are switched *on* when the repressor is removed.
Post-transcription	mRNA processing	**Splicing**: introns (non-coding DNA) are removed from mRNA. Different polypeptides can be formed by retaining some **introns** and rearranging **exons**.
	mRNA editing	mRNA can be edited by adding, deleting, or substituting nucleotides.
Translation	Control of mRNA binding	**Inhibitory proteins** prevent the binding of mRNA to ribosomes. **Initiation factors** promote mRNA binding.
Post-translation	Polypeptide modification	For example, protein folding, and the addition of non-protein groups and disulfide bridges.

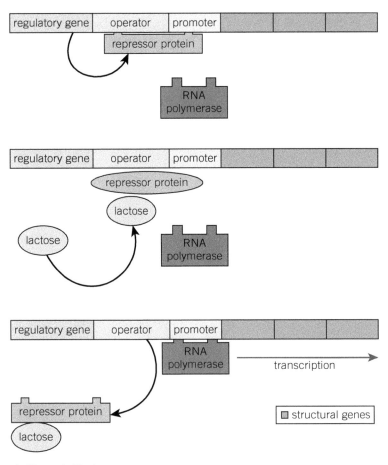

▲ **Figure 1** *The lac operon*

Synoptic link

You learned about transcription and translation in Topic 3.10, Protein synthesis.

Revision tip: Structural vs. regulatory

Regulatory genes code for proteins that control the expression of other genes (e.g. repressor proteins). Structural genes code for proteins that are not involved in regulation (i.e. most of the proteins that you have learned about, such as enzymes and hormones).

Summary questions

1 Explain how a gene consisting of 930 base pairs can be transcribed into mature mRNA consisting of only 615 nucleotides. *(2 marks)*

2 Explain the importance of splicing in organisms. *(2 marks)*

3 Figure 2 shows some different conditions shared by twins, and illustrates the relative influence of genes and the environment on each trait. Explain the evidence for the comparative influence of genetics and the environment on height and strokes. *(3 marks)*

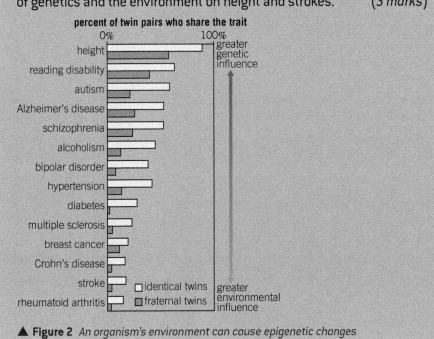

▲ **Figure 2** *An organism's environment can cause epigenetic changes to genes. This helps to explain phenotypic differences between twins*

An organism's development is governed by sets of regulatory genes. Fungi, plants, and animals all possess these DNA sequences, which are known as **homeobox genes**. The development of an organism's body plan is also reliant on the balance between **apoptosis,** otherwise known as programmed cell death, and mitosis.

Homeobox genes

Question and model answer: Homeobox genes

Q. Describe the role of homeobox genes in the development of eukaryotic organisms.

A.

- Homeobox genes are regulatory genes. Homeobox sequences within these genes are 180 base pairs long and code for polypeptide sequences that are 60 amino acids long. These polypeptides are called **homeodomains**.

- Homeodomains bind to DNA and switch genes on or off. They are transcription factors.

- Hox genes are a subset of homeobox genes that are present only in animals. Species possess different numbers of these genes (e.g. vertebrates have four clusters of Hox genes).

- Homeobox genes are expressed in a set order.

- By regulating which genes are expressed in different parts of an organism, homeobox genes control the development of the organism's body and ensure structures develop in the correct positions.

- For example, a role of homeobox genes early in development is to determine the tail and head regions of an organism (i.e. its polarity).

- Homeobox genes regulate both mitosis and apoptosis.

Apoptosis

Apoptosis is the mechanism the body employs to destroy cells in a controlled fashion. It always involves the same steps: the cell shrinks, the nucleus condenses, enzymes break down the cytoskeleton, the cell breaks into fragments held in vesicles, and macrophages digest the cell fragments.

In addition to destroying damaged cells, apoptosis is vital in development. It prunes excess cells and sculpts structures in developing organisms.

Synoptic link

You can remind yourself of how mitosis operates by reading Topic 6.2, Mitosis.

Revision tip: Mitosis vs. apoptosis

Remember that in organs and tissues both apoptosis and mitosis will be happening. The two processes are balanced if a region of tissue or an organ remains the same size. The organ will decrease in size if the rate of apoptosis exceeds mitosis.

Summary questions

1 Describe two examples of apoptosis in an organism's development.
(2 marks)

2 Suggest why species differ in the number of homeobox genes they possess.
(2 marks)

3 Necrosis is a damaging form of cell death caused by infection or trauma. Necrotic cells rupture and release hydrolytic enzymes. Outline and explain the differences between apoptosis and necrosis. *(4 marks)*

Chapter 19 Practice questions

1 What is the name given to the type of mutation that results in the
formation of a stop codon? (*1 mark*)

   A Nonsense   **C** Silent

   B Missense   **D** Insertion

2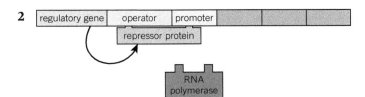

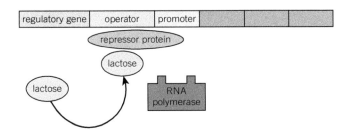

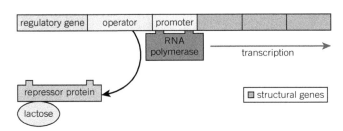

   a What is the overall name given to the DNA regions shown in the
   diagram? (*1 mark*)

   b Describe the role of lactose in the process shown in the diagram.
   (*2 marks*)

3 Complete the following passage about apoptosis by adding the most
appropriate words or phrases to the gaps. (*4 marks*)

   Apoptosis is also known as ……………………….. cell death. A cell's
   cytoskeleton is broken down by ……………………… . The cell shrinks
   and the plasma membrane forms bulges called ……………………… .
   The cell fragments and apoptotic bodies form, which are engulfed by
   ……………………… .

4 Explain how the process illustrated in the diagram below contributes to
the control of gene expression. (*3 marks*)

5 Match each description of a mutation with the correct type of
mutation. Choose from: missense, point, silent. (*3 marks*)

   A Thymine is exchanged for cytosine in the base sequence of DNA.

   B A mutation occurs in a non-coding region of DNA.

   C Three new amino acids are introduced into the primary structure of
   a protein.

20.1 Variation and inheritance

Specification reference: 6.1.2(a) and (d)

Synoptic link

You learned about the environmental and genetic causes of variation in Topic 10.5, Types of variation. Topic 10.6, Representing variation graphically, outlined the differences between continuous and discontinuous variation.

Phenotypic variation within species can have both environmental and genetic origins. The relative influence of the environment and genes determines whether a characteristic exhibits discontinuous or continuous variation.

What causes phenotypic variation?

Genetics

Members of a species can possess different versions of genes (i.e. alleles). These gene variants are produced by DNA **mutations**. Sexual reproduction shuffles alleles, producing new allele combinations in gametes through the following processes:

- crossing over (during meiosis)
- independent assortment (during meiosis)
- random fusion of gametes during fertilisation.

The different alleles in a population code for different polypeptides, which establishes phenotypic variation.

Environment

Characteristics such as eye colour and gender are unlikely to change during an individual's lifetime. However, many other traits can be sculpted by the environment. For example, body mass, while in part determined by genetics, is influenced by nutritional intake and activity levels. The length of a plant's leaves is determined by mineral intake, ambient temperatures, and its exposure to light.

Continuous vs. discontinuous variation

When both genetics and the environment play roles in shaping a characteristic, individuals exhibit continuous variation. Discontinuous variation is a result of genetics alone determining a characteristic.

Type of variation	Definition	Cause of variation	How is the data displayed?	Examples
Discontinuous	A characteristic that has specific (discrete) values (without intermediate values)	Genetics (one or two genes)	Bar graph (usually qualitative data)	Blood groups Genetic diseases (i.e. someone either has cystic fibrosis or does not)
Continuous	A characteristic that has any value within a range	Environment and genetics (usually several genes)	Line graph (quantitative data)	Body mass Height Blood glucose concentration

Revision tip: Continuous or discontinuous?

Deciding whether a characteristic shows continuous or discontinuous variation is not always straightforward. For example, we can divide eye colours into discrete categories: brown, green, blue. However, in reality, eyes come in many shades of blue, green, and brown. These colours can be measured precisely and presented in a line graph to show continuous data.

Summary questions

1 Which of the following statements represents continuous variation and which represents discontinuous variation?
    a A characteristic controlled by one gene with five different alleles.
    b Quantitative data presented in a line graph.
    c A person's lung capacity.   (3 marks)

2 A tomato plant was cloned. One cloned plant was grown at 10°C and another was grown at 20°C. Suggest and explain the effect on the size of the tomato fruit.   (3 marks)

3 Identical twins are sometimes separated early in life and are raised in different environments. Suggest how these sets of twins can be used to assess the causes of variation in human populations.   (3 marks)

20.2 Monogenic inheritance

Specification reference: 6.1.2(b)(i)

Patterns of inheritance can be illustrated with genetic diagrams, which show the probability of offspring inheriting certain alleles from their parents. Here you will learn about monogenic inheritance (i.e. inheritance patterns involving one gene), including the inheritance of codominant alleles, multiple alleles, and sex-linked genes.

Genetic crosses

Genes with one recessive allele and one dominant allele

Genetic diagrams can be drawn to illustrate the probability of inheriting a particular genotype.

 Worked example: Cystic fibrosis

Cystic fibrosis (CF) is a genetic disease with two alleles: a dominant healthy allele (F) and a faulty recessive allele (f).

Example 1: A genetic cross between a homozygous dominant father and a homozygous recessive mother.

Step 1: State the genotypes of both parents (FF and ff in this example).

Step 2: State the possible gametes that each parent could pass on, and draw a Punnett square to show the possible genotypes of their offspring.

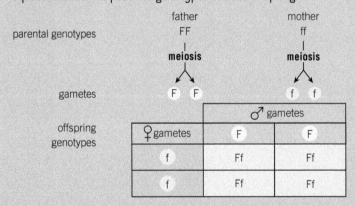

Step 3: State the probability of offspring having particular phenotypes.

In this example 100% of the offspring will be healthy carriers of the cystic fibrosis allele.

Example 2: Two heterozygous parents.

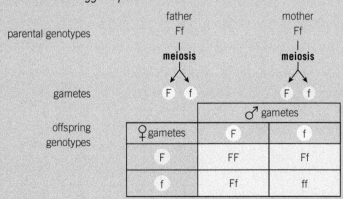

In this example there is a 25% probability that an offspring will have cystic fibrosis (genotype ff).

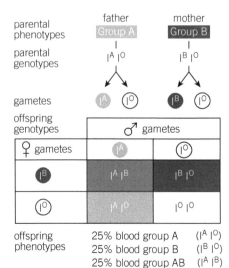

offspring genotypes	♂ gametes	
♀ gametes	I^A	I^O
I^B	$I^A I^B$	$I^B I^O$
I^O	$I^A I^O$	$I^O I^O$

offspring phenotypes		
25% blood group A	$(I^A I^O)$
25% blood group B	$(I^B I^O)$
25% blood group AB	$(I^A I^B)$
25% blood group O	$(I^O I^O)$

▲ **Figure 1** *The inheritance of the ABO blood group gene, an example of a gene with multiple alleles*

Revision tip: Representing sex linkage

When displaying sex-linked genotypes, an appropriate letter is chosen to represent the gene (e.g. 'H' for haemophilia). Lower and upper cases are used for recessive and dominant alleles. This allele letter is placed as superscript on the X or Y chromosome (e.g. X^H and X^h).

Codominance

Both alleles are expressed when a genotype consists of two codominant alleles.

For example, sickle cell anaemia is a disease in which a faulty version of the haemoglobin molecule is produced. The genotype $H^A H^A$ results in normal haemoglobin production, whereas $H^S H^S$ results in the disease. The heterozygous genotype $H^A H^S$ produces both versions of haemoglobin because the alleles show codominance.

Multiple alleles

Many genes have more than two possible versions. These multiple gene variants are a mixture of dominant, codominant, and recessive alleles. The gene for the ABO blood group, for example, has two codominant alleles (I^A and I^B) and one allele that is recessive to them both (I^O).

Sex linkage

Genes located on the X or Y chromosomes are called sex-linked. Two aspects of sex linkage are:

- more genes are found on the X chromosome because it is larger than the Y chromosome

- males are more likely to suffer from a sex-linked disease because they inherit one X chromosome. For recessive conditions, only one recessive allele would be required for them to inherit the disease.

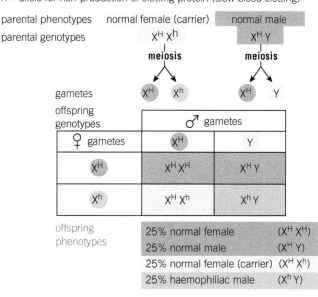

H = allele for production of clotting protein (rapid blood clotting)
h = allele for non-production of clotting protein (slow blood clotting)

offspring genotypes	♂ gametes	
♀ gametes	X^H	Y
X^H	$X^H X^H$	X^H Y
X^h	$X^H X^h$	X^h Y

offspring phenotypes		
25% normal female	$(X^H X^H)$
25% normal male	$(X^H Y)$
25% normal female (carrier)	$(X^H X^h)$
25% haemophiliac male	$(X^h Y)$

▲ **Figure 2** *Haemophilia inheritance, an example of sex linkage*

Key terms

Genotype: The genetic composition of an organism, which describes all the alleles it contains. Genotypes for a particular locus can be heterozygous or homozygous.

Phenotype: An organism's observable characteristics.

Homozygous: Having two identical alleles of a gene. Homozygous genotypes can be recessive or dominant.

Heterozygous: Having two different alleles of a gene.

Summary questions

1. Suggest why more X-linked traits exist than Y-linked traits. *(2 marks)*

2. The colour of the four o'clock flower (*Mirabilis jalapa*) is controlled by two codominant alleles: C^R (red flowers) and C^W (white flowers). What are the genotypes of the two plants that would need to be bred to produce 50% red offspring and 50% pink offspring? *(2 marks)*

3. Paul and Sandra have a child together. Neither has cystic fibrosis, but Sandra is a carrier (genotype Ff). Paul has the genotype FF. Their child, many years later, has a baby whose other parent is a carrier of cystic fibrosis. Calculate the probability that Paul and Sandra will have: **a** a child with cystic fibrosis **b** a child who is a carrier of cystic fibrosis **c** a grandchild with cystic fibrosis **d** a grandchild who is a carrier of cystic fibrosis. *(4 marks)*

20.3 Dihybrid inheritance

A dihybrid cross shows the inheritance pattern of two characteristics, coded for by two different genes.

Genetic cross with two genes

Genetic diagrams can be drawn to illustrate dihybrid inheritance. These diagrams have a similar format to the monohybrid crosses in Topic 20.2, Monogenic inheritance, but at least four possible alleles exist (two per gene) and therefore four potential phenotypes exist.

 Worked example: *Drosophila* eyes and wings

Two observable traits in *Drosophila* flies are eye colour and wing size. The allele for red eyes is R (dominant). The recessive allele for white eyes is r (recessive). Wing shapes can be normal (W, a dominant allele) or vestigial (w, a recessive allele). The possible genotypes of offspring from two heterozygous parents (RrWw) are shown in the genetic diagram below.

Male gametes	Female gametes			
	RW	Rw	rW	rw
RW	RRWW	RRWw	RrWW	RrWw
Rw	RRWw	RRww	RrWw	Rrww
rW	RrWW	RrWw	rrWW	rrWw
rw	RrWw	Rrww	rrWw	rrww

The phenotypes of offspring will be present in the following ratios:

9 red eyes, normal wings

3 white eyes, normal wings

3 red eyes, vestigial wings

1 white eyes, vestigial wings

A scientist can assess whether their predictions about the genetics of a species are correct by using the chi-squared test, which you will learn about in Topic 20.4, Phenotypic ratios. If the phenotypic ratio for a dihybrid cross differs from what is expected then other effects, such as epistasis and linkage, might be present.

Revision tip: Phenotypic ratios

Keep an eye out for certain phenotypic ratios in offspring. For example, if two heterozygous parents are used for a dihybrid cross then the expected offspring phenotypes will be in a 9:3:3:1 ratio.

Summary questions

1. A white-eyed male *Drosophila* with vestigial wings had offspring that were all red-eyed with normal wings. Predict the genotype of the mother. *(2 marks)*

2. A scientist studied the inheritance of two traits in pea plants: seed colour and seed texture. The alleles for these traits are: R = round seeds, r = wrinkled seeds, G = yellow seeds, g = green seeds. R and G are the dominant alleles. What proportion of each phenotype would you expect to find in the offspring of a plant with the genotype RRGg and another plant that is RrGg? *(2 marks)*

3. When predicting a phenotypic ratio of 9:3:3:1 for a dihybrid cross, what assumptions do you have to make about the two gene loci being studied? *(2 marks)*

20.4 Phenotypic ratios

Specification reference: 6.1.2(b)(ii) and (c)

Simple inheritance patterns can be predicted when two genes are inherited independently. Sometimes, however, two genes are connected in some way. The **chi-squared** (χ^2) **test** compares actual offspring phenotypes against the expected phenotypic ratios. If the offspring phenotypic ratios differ from what is expected then this suggests the genes are not inherited independently. For example, two genes can be located on the same chromosome (**autosomal linkage**), or one gene can affect the expression of another (**epistasis**).

Autosomal linkage

Two genes are said to be linked when they are located on the same autosome (i.e. a chromosome other than X or Y). This is important because:

- Linked allele combinations will be inherited together (as a single unit).
- Only crossing over during meiosis (see Topic 6.3, Meiosis) can separate linked allele combinations.
- The nearer two genes are on a chromosome, the less likely they are to be separated during crossing over.

Synoptic link

You learned about sex linkage in Topic 20.3, Dihybrid inheritance. Do not confuse this with autosomal linkage.

Epistasis

Epistasis occurs when one gene prevents the expression of another gene. Two forms of epistasis exist: recessive and dominant.

	Description	Example
Recessive epistasis	The epistatic gene (i.e. the gene doing the suppressing) needs to be **homozygous recessive** to prevent the expression of the other gene.	Flower colour in *Salvia* is controlled by two genes. Dominant allele B = purple flowers Recessive allele b = pink flowers However, when another gene is homozygous recessive (aa), the flowers are white. The genotype of B/b becomes irrelevant. AaBB, AABB, AABb, AaBb = purple flowers Aabb, AAbb = pink flowers aaBB, aaBb, aabb = white flowers
Dominant epistasis	The epistatic gene needs **at least one dominant allele** to prevent the expression of the other gene.	Fruit colour in summer squash is also controlled by two genes. Dominant allele E = yellow fruit Recessive allele e = green fruit However, the presence of one dominant D allele at another gene results in white fruit. ddEe, ddEE = yellow fruit ddee = green fruit DdEE, DdEe, Ddee, DDEE, DDEe, DDee = white fruit

Chi-squared test

The χ^2 test assesses whether there is a significant difference between the observed and expected numbers of offspring phenotypes.

$$\chi^2 = \Sigma \frac{(O-E)^2}{E} = \text{the sum of } \frac{(\text{observed numbers} - \text{expected numbers})^2}{\text{expected numbers}}$$

Model question and answer: Calculating chi-squared

The results from the *Drosophila* genetic cross shown in Topic 20.3, Dihybrid inheritance, can be analysed using the χ^2 test.

Q. Calculate χ^2 for this genetic cross and determine whether the predicted model of inheritance can be accepted.

> The expected values are based on the predicted phenotypic ratios of the offspring (see Topic 20.3, Dihybrid inheritance).

A.

Phenotype	Observed (O)	Expected (E)	O – E	(O – E)²	(O – E)² / E
Red eyes, normal wings	87	90	–3	9	0.100
White eyes, normal wings	31	30	1	1	0.033
Red eyes, vestigial wings	35	30	5	25	0.833
White eyes, vestigial wings	7	10	–3	9	0.900
					$\chi^2 = 1.87$

The next stage is to compare the calculated value to a χ^2 significance table.

The degrees of freedom are calculated as the number of categories (phenotypes) – 1. In this example we have 3 degrees of freedom.

The *P* value tells us the probability of differences being a result of chance. We say that differences are statistically significant if $P = 0.05$ or less (i.e. $P = 0.05$ would mean there is only a 5% probability that differences can be attributed to chance). In this case, the critical value of χ^2 (which represents a *P* value of 0.05) would be 7.81.

In our example, the P value for a calculated χ^2 **of 1.87** lies somewhere between 0.5 and 0.9. We can be very confident that the differences between our observed and expected results are down to chance and are not significant. This means we can accept our model of inheritance for these two traits.

Key terms

Autosomal linkage: Genes located on the same (non-sex) chromosome.

Epistasis: The effect of one gene on the expression of another.

Revision tip: More phenotypic ratios

Remember that a 9:3:3:1 offspring phenotypic ratio is expected if both parents are heterozygous for two unconnected genes.

Linkage will alter this ratio. Many different ratios can occur, depending on the alleles that are linked and how closely they are linked.

Dominant epistasis is likely to produce a 12:3:1 or 13:3 ratio (with heterozygous parents).

Recessive epistasis will produce a 9:3:4 offspring phenotypic ratio (with heterozygous parents).

Summary questions

1 Humans with red hair tend to have pale skin and green eyes. Suggest a reason for this inheritance pattern. *(2 marks)*

2 Pea plants have the following alleles for seed colour and texture: R = round, r = wrinkled, G = yellow, g = green.
Two pea plants were bred together to produce 100 offspring. A scientist expected 50 plants that produced round, yellow seeds and 50 that produced wrinkled, yellow seeds. She expected no green seeds. The results were: 55 round, yellow seeds; 45 wrinkled, yellow seeds; 0 round, green seeds; 0 wrinkled, green seeds. Calculate whether the scientist's predictions can be supported by the test. The critical value of χ^2 in this case is 3.84. Suggest the likely genotypes of the two plants being bred. *(4 marks)*

3 Suggest a likely molecular mechanism for
   a recessive epistasis
   b dominant epistasis. *(4 marks)*

20.5 Evolution

Evolution occurs when genetic mutations and natural selection result in a change of allele frequencies in a population. In theory, allele frequencies remain constant in a stable population. These allele frequencies can be calculated using Hardy–Weinberg equations, which are discussed here. When a population decreases in size, however, allele frequencies are liable to change.

Factors affecting allele frequencies
Forms of natural selection

A selection pressure can alter the distribution of phenotypes in a population. The selection of phenotypes can be stabilising or directional.

Form of selection	Description	Appearance (arrows show selection pressure)	Example
Stabilising	Selection favours **average phenotypes**. Alleles that produce extreme traits are eliminated.	evolved population original population	Most mammals will have fur length close to the mean in an environment with a stable temperature. Individuals with short or long fur are less likely to survive and reproduce. These alleles are therefore eliminated.
Directional	Occurs when **environmental conditions change**. Selection favours individuals with **extreme phenotypes**.		Mean fur length will increase if the mean temperature decreases in an environment. Individuals with longer fur (i.e. those with more extreme phenotypes) will have higher survival rates. Similarly, mean fur length will decrease if environmental temperature increases.

original population has eight different alleles occurring at various frequencies

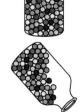

chance event reduces the size of the population significantly

the individuals that survive have fewer alleles (just four types) and with different frequencies

as the population recovers, the number and frequency of the alleles are the same as those of the population that came through the bottleneck. This population is less diverse than the original population

▲ **Figure 1** *An illustration of a genetic bottleneck effect*

Genetic drift

Genetic drift is a random change in allele frequencies. Its effects are more noticeable in small populations. For example, a **genetic bottleneck** is a drastic reduction in population numbers (e.g. due to natural disaster or environmental change). The proportions of alleles in the surviving population could be very different to those in the original population.

The **founder effect** can be considered a type of genetic bottleneck. This occurs when a small group breaks away from the original large population to form a new colony (e.g. a group of birds migrating to a new island).

The Hardy–Weinberg equations

The Hardy–Weinberg equations are used to calculate the proportions of alleles and genotypes in a population. Calculations are carried out for genes with two variants (one dominant and one recessive allele). The first equation we need to understand is:

$$p + q = 1.0$$

Where p = the frequency of the dominant allele, and q = the frequency of the recessive allele. If only two alleles exist, their frequencies must add up to 1.0 (100%).

The second equation considers the frequencies of the possible genotypes in a population:

$$p^2 + 2pq + q^2 = 1.0$$

p^2 = the homozygous dominant genotype; $2pq$ = the heterozygous genotype; q^2 = the homozygous recessive genotype. The frequencies of the three possible genotypes must add up to 1.0 (100%) since no other genotypes are possible.

Worked example of Hardy–Weinberg calculations: Cystic fibrosis

Ireland has the world's highest rate of cystic fibrosis (CF). A recent study estimated that 1 in 1353 births are of babies with CF. From this information alone, how do we calculate the proportion of unaffected carriers of the CF allele in Ireland's population?

1 CF is caused by a recessive allele (f). Only homozygous recessive genotypes produce CF.

2 The homozygous recessive genotype = ff = q^2 in the Hardy–Weinberg equation.

3 q^2 = 1 in 1353 = 0.000 739 098

4 To calculate the frequency of the recessive allele (q) we need to find the square root of q^2. $\sqrt{0.000739098}$ = 0.027 186 362

5 $p + q = 1.0$, so $p = 1 - q$. We now know $q = 0.027$ 186, therefore $p = 1 - 0.027$ 186 = 0.972 814.

6 We are now able to calculate the frequency of heterozygous genotypes (Ff) in Ireland. Ff is represented in the Hardy–Weinberg equation by $2pq$. $2pq = 2 \times 0.972$ 814 $\times 0.027$ 186 = 0.052 894.

7 A frequency of 0.052 894 means 5.3% of Ireland's population are carriers of the CF allele. Another way of expressing the same value is that approximately 1 in 19 of the population are carriers.

Synoptic link

You learned about evolution and natural selection in Topics 10.4, Evidence for evolution, and 10.8, Changing population characteristics.

Revision tip: Making assumptions

The Hardy–Weinberg equations assume that the following conditions are present in a population: no new mutations, no migration, no natural selection for or against alleles, a large population, and random mating.

In reality, these conditions rarely exist. The Hardy–Weinberg principle nonetheless provides a basis for the study of gene frequencies.

Summary questions

1 Explain how genetic bottlenecks decrease genetic diversity. *(2 marks)*

2 Albinism is a condition in which people lack melanin pigment in their hair, skin, and eyes. It is caused by a recessive allele, which means only homozygous recessive genotypes produce albinism. 1 in 17 000 people worldwide have the condition. Calculate the allele frequency of the recessive allele for albinism. *(2 marks)*

3 Ellis–van Creveld (EVC) syndrome is a disorder that includes symptoms such as dwarfism and extra fingers. The allele that causes the disease is recessive. In one Amish population, 5 in 1000 people have EVC. Calculate the percentage of the Amish population that carries one copy of the EVC allele. 1 in 123 people in the general population carry one EVC allele.
Calculate the difference in the percentage of EVC carriers between the two populations. *(5 marks)*

20.6 Speciation and artificial selection

New species can evolve through natural selection. The process of artificial selection requires human populations to provide a selection pressure by choosing animals and plants with desirable traits.

Speciation

For a new species to evolve, the following events must occur:

- Members of a population become **isolated** from the rest of the population (which prevents gene flow between the two groups).
- Genetic **mutations** continue to occur in both groups, producing **new alleles**.
- The different groups experience different **selection pressures**.
- Different alleles are selected in the two groups.
- Over generations, the two groups become genetically different and **unable to reproduce fertile offspring** together.

Two mechanisms of speciation can occur: **sympatric** and **allopatric** speciation.

Type of speciation	Description	Form of reproductive isolation	Examples
Allopatric	Members of a population are separated by a physical barrier	Geographical	Migration to different islands. A mountain range. Agricultural activity.
Sympatric	Speciation occurs within a population that shares the same habitat. This is very rare, and the theory is controversial.	Temporal Behavioural Mechanical	Differences in the timing of flowering. Different mating rituals/calls. Incompatible reproductive systems.

Artificial selection

By choosing which organisms to breed together over many generations, humans have accelerated the process of evolution. For example, tameness has been selected in cats and dogs, and crop plants have been formed with high yields. This process of artificial selection (**selective breeding**) can cause problems, however, because it requires **inbreeding** (breeding closely related individuals):

- Genetic diversity is reduced.
- Inbred populations are less able to adapt to changing environmental conditions.
- Homozygous recessive disorders are more likely to occur.

> **Revision tip: Defining a species**
> Scientists have long debated the most appropriate definition of a species. A popular definition is 'a group of individuals that can breed to produce fertile offspring'. However, this definition excludes asexual organisms, such as bacteria. An alternative definition (the phylogenetic species concept) is 'a group of organisms sharing similar morphology, genetics, biochemistry, and behaviour'.

Summary questions

1 Suggest the type of reproductive isolating mechanism that exists in the following examples:
   a The Kaibab squirrel and the Abert squirrel are separated by the Colorado river.
   b The white sage plant has a large landing platform for pollinating insects, whereas the black sage has a small landing platform for pollinators.   (2 marks)

2 Apple maggot flies lay eggs on either hawthorns or domestic apples. Females tend to lay their eggs on the type of fruit they grew up in, and males usually search for mates on the type of fruit they grew up in. Suggest how speciation might arise in this species.   (2 marks)

3 Read the following passage and evaluate the statement 'bonobos and chimpanzees should be considered members of the same species'.   (2 marks)
   Bonobos and chimpanzees may have become separated and geographically isolated by the Congo River. However, bonobos and chimpanzees will mate in captivity, and their hybrid offspring are thought to be fertile. Scientists have estimated that more than 1 million years on average is required for reproductive isolation in primates. Chimpanzees and bonobos probably diverged roughly 1 million years ago. The behavioural and anatomical differences between them, which have evolved during their geographical isolation, are insufficient to stop them mating. Nor have post-mating barriers evolved to render any hybrid offspring infertile.

1 Which of the following statements is/are true of the founder effect?

(1 mark)

1 It can be considered a form of genetic bottleneck.

2 It often results in a change in allele proportions within a population.

3 It is usually a result of two populations merging.

A 1, 2, and 3 are correct

B Only 1 and 2 are correct

C Only 2 and 3 are correct

D Only 1 is correct

2 Which of the following statements is/are true of the assumptions made when applying the Hardy–Weinberg principle to calculate the frequency of alleles in a population? *(1 mark)*

1 Natural selection is operating.

2 The population is large.

3 Mating is random.

A 1, 2, and 3 are correct

B Only 1 and 2 are correct

C Only 2 and 3 are correct

D Only 1 is correct

3 Two parents are both carriers of the recessive allele (f) that causes a rare condition called Friedreich's ataxia (FA).

The parental genotypes are Ff. People with FA have the genotype ff.

The two parents have a daughter. Their daughter, later in life, has a child with a father who is a carrier of FA. What is the probability that this child (the grandchild of the first couple) has FA? *(1 mark)*

A 0%   C 50%

B 25%   D 75%

4 Which of the following statements is/are true of recessive epistasis?   *(1 mark)*

1 Two gene loci are involved.

2 A homozygous recessive genotype is required for the expression of another gene.

3 AaBb would produce an epistatic effect.

A 1, 2, and 3 are correct

B Only 1 and 2 are correct

C Only 2 and 3 are correct

D Only 1 is correct

5 Which of the following statements is an example of temporal isolation? *(1 mark)*

A Separation by a river.

B Differences in mating displays.

C Differences in breeding season.

D Reproductive system incompatibility.

6 Outline, with examples, how the founder effect can change allele frequencies in populations. *(6 marks)*

A DNA profile is a genetic fingerprint that is unique to each person (except identical twins). DNA profiles are constructed using techniques such as the polymerase chain reaction (**PCR**) and **electrophoresis**.

Producing a DNA profile

To produce a DNA profile, DNA must be:

- **extracted** (and many **copies made** using **PCR**)
- **digested** (broken into fragments) using **restriction endonucleases**
- **separated** using **electrophoresis**
- **hybridised** with **probes** (which bind to fragments and enable them to be visualised)
- **visualised** in banding patterns (bars).

Techniques used in DNA profiling

Technique	Purpose	The process
PCR	DNA amplification (**copying**)	The DNA sample is placed in a thermocycler (which cycles through three temperatures) **95°C**: breaks hydrogen bonds in the DNA, splitting it into two strands **55°C**: primers bond to the end of each DNA strand **72°C**: Taq DNA polymerase joins free nucleotides to each strand.
Electrophoresis	**Separation** of DNA fragments	DNA fragments are placed at the end of a gel plate. A positive electrode is at the opposite end of the plate. DNA moves towards the positive electrode when a current is applied (because all DNA fragments have phosphate groups with negative charges). Longer fragments move slower, shorter fragments move faster. The DNA fragments are therefore separated into bands based on size.

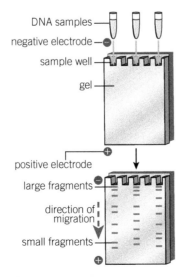

▲ **Figure 1** *Electrophoresis*

Synoptic link

You learned about DNA structure and replication in Topics 3.8 and 3.9.

Maths skill: log scales in PCR

The amplification of DNA is an example of exponential increase. Log scales can be used to show the relationship between cycles of heating and cooling and the increases in copy number.

$$1024 = 2^{10}$$

After 10 cycles, a single fragment of DNA can, in theory, be amplified to 1024 fragments, which would be represented on a log scale as $10^{3.01}$.

Plotting the PCR cycle number against the amount of DNA on a log scale produces a straight line on a graph.

Revision tip: Restriction endonucleases

Restriction endonucleases are enzymes found in bacteria. More than 50 of these enzymes are known, and each one cuts DNA at specific base sequences (recognition sites). As well as digesting DNA prior to electrophoresis, they are used for genetic engineering (see Topic 21.3, Using DNA sequencing).

Revision tip: Taq DNA polymerase

Human DNA polymerase is not used in PCR. Instead, the enzyme is obtained from a thermophilic bacterium, *Thermus aquaticus* (Taq). This form of polymerase is tolerant to heat so does not denature during temperature cycling.

Revision tip: VNTRs

DNA profiles are often formed from sections of DNA called **variable number tandem repeats** (VNTRs). The patterns of VNTRs differ between people; therefore the probability of two unrelated individuals having the same VNTR profile is very low.

Revision tip: Uses of profiling

DNA profiles can be used for:

Forensics (legal applications), such as in criminal investigations and paternity testing.

Disease risk analysis

Classification

Summary questions

1   You learned about chromatography in Topic 3.6, Structure of proteins. Suggest why electrophoresis is sometimes described as 'similar to chromatography'. *(2 marks)*

2   If PCR begins with one fragment, calculate the number of fragments (represented on a log scale) produced after
    a   5 cycles
    b   12 cycles
    c   17 cycles. *(3 marks)*

3   Describe two similarities and two differences between semi-conservative replication of DNA and PCR. *(4 marks)*

21.2 DNA sequencing and analysis
21.3 Using DNA sequencing

Specification reference: 6.1.3(a) and (b)

Scientists are now able to work out the base sequences of genes, which has improved their ability to classify organisms and assess the risk of disease. DNA sequencing also raises the possibility of synthetic biology: the redesign of genes and biological systems.

How is DNA sequenced?

- PCR is conducted (see Topic 21.1, DNA profiling).
- However, some of the free nucleotides in PCR have been modified in two ways:
  - When they bond to a DNA strand they terminate polymerisation.
  - They are fluorescently coloured – A, T, C, and G have different colours.
- New DNA strands stop growing whenever a terminator base is added – PCR is interrupted.
- This results in every possible chain length being produced (e.g. if the full DNA sequence has 1000 bases, chain lengths of 999, 998, 997, etc. will be generated).
- Lasers detect the final base on each chain.
- The sequence of DNA bases can therefore be worked out.

Figure 1 *DNA is sequenced by producing different chain lengths with terminating bases that are fluorescently tagged*

How are DNA sequences used?

Use	Description
Disease analysis	Particular gene variants can be sequenced and linked to the risk of inheriting certain diseases. Sequencing pathogen genomes enables identification of antibiotic-resistant bacteria, pinpointing genetic markers for vaccines, and identification of targets for drugs.
Classification	Identifying species by using DNA barcodes. Studying evolutionary relationships by comparing similarities and differences between species' base sequences.
Genotype– phenotype relationships	Amino acid sequences do not always match those predicted from base sequences – several phenotypes are possible from the same genotype. Knowledge of both amino acid and base sequences enables comparisons to be made.
Synthetic biology	Genetic engineering (see Topic 21.4) requires knowledge of base sequences.

Common misconception: Bioinformatics

Bioinformatics and *computational biology* are often used interchangeably. However, the terms are subtly different. Bioinformatics is a toolkit – the creation of databases and computer software that can be used to solve biological questions. Computational biology is the application of bioinformatics. For example, the sequencing of genomes relies on bioinformatics and is therefore an example of computational biology.

 Go further: Advances in sequencing

The original DNA sequencing method used interrupted PCR in capillaries. Several improvements to the process have been introduced, including:

- The process can now take place on a plastic slide, called a flow cell; millions of DNA fragments can be attached to the slide.

- The use of reversible terminator bases enables a high yield of sequenced DNA.

- The fluorescent tags can be visualised at the same time.

- Pyrosequencing: the binding order of the four bases to a template DNA strand is monitored by the release of light from the luciferase enzyme.

  Many additional sequencing methods are being developed.

1 Suggest what criteria are used to judge the effectiveness of a sequencing method.

Revision tip: Profiles vs. sequences
Do not confuse DNA profiling (producing a genetic fingerprint, unique to a person, which is based on particular sections of DNA) and sequencing (determining the precise base sequences in DNA).

Summary questions

1 Explain the use of sequencing in
   a the analysis of disease risk.
   b classification. *(4 marks)*

2 Outline the differences and similarities between the use of PCR for amplification and PCR for sequencing. *(4 marks)*

3 Suggest why protein amino acid sequences sometimes differ from those predicted from DNA sequences. *(3 marks)*

21.4 Genetic engineering

Specification reference: 6.1.3(f)

Knowledge of base sequences opens up the possibility of altering genes and transferring them between species. This is known as genetic engineering.

Genetic modification techniques

Obtaining the desired gene

A desired gene can be extracted using two methods: producing the gene from an mRNA template, or cutting out the gene using restriction endonucleases.

Enzyme used for gene extraction	Key points
Restriction endonucleases	The gene is cut from its source DNA. Cutting the gene and plasmid with the same enzyme produces two sets of **sticky ends** with complementary base pairings, enabling the gene to be inserted into the plasmid.
Reverse transcriptase	mRNA (transcribed from the desired gene) is extracted from cells. Reverse transcriptase is used to convert mRNA to cDNA (a single strand of complementary DNA).

Using vectors to produce recombinant DNA

A vector is used to transfer the gene into the organism that is being modified. Vectors include:

- **Plasmids**: circular bacterial DNA; by far the most common vector. Bacterial artificial chromosomes (**BAC**s) are synthetic structures based on plasmids. DNA **ligase** is used to seal the gene into the plasmid.

- **Viruses** (e.g. **bacteriophages**, which naturally infect bacterial cells).

The vector, containing the donated gene, must then be transferred into the recipient's cells. This process is called **transformation**. Several methods of transformation are used, as outlined in the table below.

Method of vector transfer	For which organisms is this method used?	Key points
Culture heating	Bacteria	Bacterial cell membranes become more permeable when heated in a calcium-rich solution. Plasmids are then able to diffuse into the cells.
Electroporation	Bacteria and unicellular eukaryotes	Electric current disrupts the cell membrane, enabling plasmids to enter.
Electrofusion	Plants	Electric currents enable the cell and nuclear membranes of two different cells to fuse.
Viral transfer	Plants, bacteria, and animals	Viruses naturally infect cells, and this mechanism can be exploited to insert DNA directly into target cells.
Agrobacterium tumefaciens infection	Plants	*A. tumefaciens* naturally infects plant cells and can be used to introduce recombinant plasmids.

Key terms

Recombinant DNA: DNA combined from two different sources.

Plasmid: Small circular piece of DNA found in prokaryotic cells.

Revision tip: Sticky notes

Many restriction endonucleases produce a staggered cut when digesting DNA. This creates two short sequences of exposed, unpaired bases known as **sticky ends**.

Revision tip: Marker genes

Marker genes (sometimes called reporter genes) indicate whether or not a gene (and therefore a plasmid) has been successfully taken up by bacterial cells. Marker genes are usually for **antibiotic resistance** or **fluorescence**.

Summary questions

1 Explain why, in genetic engineering, it is important to use the same restriction endonuclease to cut both the gene being transferred and the plasmid vector. (*2 marks*)

2 Suggest the advantages of genetically modifying a crop to be herbicide resistant. (*2 marks*)

3 Knockout studies can inactivate a gene by replacing the functional version with a non-functional artificial DNA sequence in a study animal (e.g. mice). Explain how a scientist could use this technology to test the association between a particular gene and the risk of cancer. (*2 marks*)

21.5 Gene technology and ethics

Specification reference: 6.1.3(g) and (h)

Gene technology is a fast-developing field. Some people worry that ethical considerations are being overlooked as the science progresses.

Genetic engineering ethics

Type of engineering	What are the potential ethical issues?
Genetically modified (GM) microorganisms	The use of the technology for biological warfare.
Pest resistance in plants	GM plants could produce toxins that might harm insects other than the targeted pests.
Pharming (producing human medicines from GM animals)	Is animal welfare compromised? Will genetic engineering damage the health of animals?
Patenting (i.e. legal ownership of GM technology)	How available will the technology be for those it might benefit? Companies are able to patent techniques and GM organisms, and farmers in poor countries are forced to pay for their use.

Gene therapy

Gene therapy is the addition of beneficial alleles to the cells of people with disease-causing alleles.

Type of gene therapy	Which cells are targeted?	Examples	Limitations
Somatic cell therapy	Human body cells	Haemophilia, cystic fibrosis, immune diseases	A possible risk of additional health problems. Introduced genes are non-heritable. Requires repeat treatments.
Germline therapy	Gametes or early embryonic cells	None – illegal in humans	Ethical issues – changes are permanent, and who decides which genes are targeted?

Revision tip: Gene therapy transfer methods
Genes can be transferred into target cells using either harmless **viruses** or **liposomes** (hollow spheres of lipid molecules).

Summary questions

1 Why does gene therapy not provide a permanent cure for cystic fibrosis? *(2 marks)*

2 Why is gene therapy potentially most useful for treating diseases stemming from single-gene mutations? *(2 marks)*

3 Suggest why viruses and liposomes are used as vectors for delivering genes to target cells. *(2 marks)*

1 During the polymerase chain reaction, why is the temperature lowered to 55°C? *(1 mark)*

   A To enable DNA polymerase to function.

   B To enable primers to attach to the DNA strands.

   C To separate DNA strands.

   D To enable free phosphorylated nucleotides to attach to DNA strands.

2 After 12 cycles of PCR, a single fragment of DNA can, in theory, be amplified to how many fragments? *(1 mark)*

   A $10^{3.01}$

   B $10^{3.31}$

   C $10^{3.61}$

   D $10^{3.91}$

3 Which of the following statements is/are true of the sticky ends produced during genetic modification of bacteria? *(1 mark)*

   1 They are often palindromic.

   2 They are produced by DNA ligase.

   3 They are usually more than 20 bases in length.

   A 1, 2, and 3 are correct

   B Only 1 and 2 are correct

   C Only 2 and 3 are correct

   D Only 1 is correct

4 Identify the following molecules that are used in genetic engineering:

   A An enzyme that catalyses the production of new DNA.

   B Several enzymes that can cut DNA at specific sequences.

   C An enzyme that anneals (seals) sticky ends.

   D Circular DNA (often containing antibiotic resistance genes) found in bacteria.

   E An enzyme that converts RNA to DNA. *(5 marks)*

5 Four scientific aims are described below.

   A Generating genetically identical transgenic mice that show disease symptoms.

   B The production of a genetic fingerprint unique to an individual.

   C Transformation using a recombinant plasmid.

   D Genetic manipulation of adult cells to cure a genetic disease.

   Match the correct letters to the following procedures:

   Genetic engineering.

   Somatic cell therapy

   DNA profiling

   Animal reproductive cloning *(4 marks)*

22.1 Natural cloning in plants
22.2 Artificial cloning in plants

Specification reference: 6.2.1(a) and (b)

Many plant species are able to reproduce asexually to form clones. Humans can exploit this cloning potential to produce large numbers of genetically identical plants for commercial benefit.

Natural cloning

Plants form natural clones through a process called **vegetative propagation**. The clones are produced via mitosis from undifferentiated **meristem** cells. Depending on the species, vegetative propagation can be initiated from roots, shoots, or leaves.

> ### Revision tip: How are plants able to clone themselves?
> The ability to form natural clones is much more common in plants than animals. This is because many plant cells are undifferentiated and **totipotent** (i.e. retain the potential to divide into a whole organism – see Topic 6.5, Stem cells). Totipotent **stem cells** are located in regions of plants called **meristem** tissue (found principally in roots, shoots, and cambium).

▼ **Table 1** *Examples of vegetative propagation forming natural clones in plants*

Type of vegetative propagation	Example	Description
Root suckers	Elm trees	Root suckers/basal sprouts grow from meristem cells in the tree trunk close to the ground.
Runners	Strawberry plants	Stems grow sideways along the ground from the parent plant. Roots develop where the runner touches the ground.
Tubers	Potatoes	Underground stems swell with nutrients and develop into new plants.
Bulbs	Daffodils	Leaf bases swell with nutrients, and buds develop into new plants.
Rhizomes	Marram grass	Underground stems develop buds and form new vertical shoots.

Practical skill: Plant cuttings

Humans can produce clones of plants by cutting and replanting stems. This exploits and speeds up plants' natural cloning processes. The key aspects of growing cuttings are:

- Short sections of stems are cut and planted.

- Rooting hormone is applied to encourage root growth.

- The cutting is watered thoroughly and covered with a plastic bag for several days to create warm, moist conditions.

Artificial cloning

Humans can produce many genetically identical plants through **micropropagation**. This is especially useful when the desired plant is rare, produces few seeds, or does not readily produce natural clones.

Practical skill: Micropropagation

The key aspects of micropropagation are:

- A small sample of meristem tissue is taken from shoot tips. The tissue that is removed is called an **explant**.

- The explant is **sterilised** (e.g. in ethanol or sodium dichloroisocyanurate).

- The sterilised explant is placed in a **culture** medium (i.e. a solution containing an ideal balance of **nutrients** and plant **hormones**).

- The explant cells divide to form a mass of undifferentiated cells called a **callus**.

- The callus is transferred to a new culture medium, which contains hormones that encourage differentiation and shoot growth.

- The developing plantlets are transferred into soil.

Synoptic link

You learned about the roles of various plant hormones in Chapter 16, Plant responses.

Pros and cons of artificial cloning

▼ **Table 2** *Advantages and disadvantages of artificial cloning by micropropagation*

Advantages	Disadvantages
Rapid production. **Seedless**, sterile crop plants (e.g. seedless grapes) can be produced. The genetic make-up of the propagated plants is known and **desired traits** can be retained in the clones.	Quite **expensive**. The **lack of genetic variation** means all the clones are vulnerable to the same diseases or environmental change.

Summary questions

1 Describe how the presence of meristem tissue in many plant species enables them to form natural clones. *(3 marks)*

2 Plant hormones are added to cultures during micropropagation. A balanced ratio of auxin to cytokinins promotes callus formation. Suggest why auxin concentration might be raised above cytokinin concentration at certain stages of micropropagation. *(2 marks)*

3 Explain why vegetative propagation is likely to be more effective at enabling plants to recover from a fire than a pathogenic infection. *(2 marks)*

Key terms

Vegetative propagation: Asexual reproduction in plants to produce clones.

Micropropagation: The use of tissue culture to produce many artificial clones of a plant.

Clones: Genetically identical individuals that result from asexual reproduction.

Adults of some animal species can produce clones, although this is less common than cloning in plants. Scientists have developed techniques to produce clones of commercially beneficial animals.

Natural cloning

Many invertebrate animals produce clones (e.g. species of *Hydra* generate buds that develop into clones, and starfish form from fragments of an original animal). Very few vertebrates are known to produce clones of themselves. The endangered sawfish is the first vertebrate species to be observed successfully cloning in the wild.

> **Revision tip: For the twin!**
> **Monozygotic twins** are produced when an embryo splits at an early stage of development to form two separate embryos. The two resultant offspring originate from the same zygote and are clones of each other (but are not clones of either parent).

Go further: Aphids

Aphids (sometimes called greenflies) are insects. Some species of aphid can reproduce both sexually and asexually, depending on the conditions in their environment. During spring and summer, asexual reproduction generates clones of female aphids; this process is called **parthenogenesis** and offspring are born as nymphs without the need for eggs. During the autumn and early winter, sexual reproduction produces eggs that hatch the following spring.

1 Suggest what benefit aphids gain from using asexual reproduction during the summer.

2 Suggest why aphids switch to sexual reproduction prior to the winter.

Artificial cloning

Scientists can now produce clones of vertebrates using two techniques: **embryo splitting** (which mimics the natural twinning process) and **somatic cell nuclear transfer**.

▼ **Table 1** *The two methods for artificially cloning animals*

	Embryo splitting/artificial twinning	Somatic cell nuclear transfer (SCNT)
What is done?	Sperm from a male with desired traits is used to fertilise eggs from a desired female (via **artificial insemination** or *in vitro* fertilisation). An **embryo is split** into several smaller embryos and grown in a lab. Each embryo is implanted into a **surrogate mother**.	The **nucleus** from an adult somatic (body) cell is transferred into an **enucleated egg** (an egg lacking its own nucleus) using **electrofusion**. The resultant embryo is transferred into a **surrogate mother**.
Nature of the clones	Offspring are clones of each other, but (like regular offspring) share 50% of their alleles with the father and 50% with the mother.	Offspring are clones of the original body cell (from which the nucleus was taken).

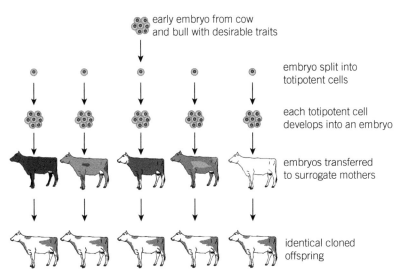

▲ **Figure 1** *Artificial cloning by embryo splitting*

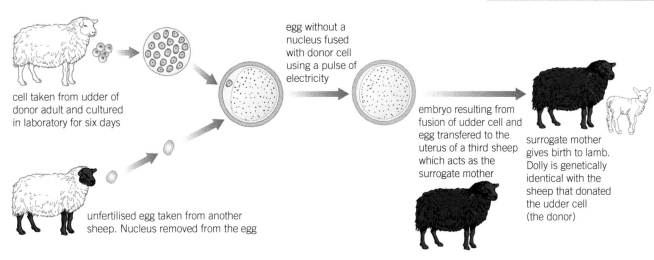

▲ **Figure 2** *Artificial cloning by somatic cell nuclear transfer*

▼ **Table 2** *The pros and cons of artificial animal cloning*

Pros of animal cloning	Cons of animal cloning
Embryo splitting enables **many more offspring** to be produced from the best farm animals. SCNT has the potential to reproduce specific animals, such as pets, and rare or extinct animals.	SCNT is very **inefficient** (i.e. high failure rate). Cloned animals may have shortened lifespans or **health problems**.

Summary questions

1 Describe the difference between reproductive and non-reproductive cloning. *(2 marks)*

2 Explain why an embryo should be split at an early stage for artificial twinning to succeed. *(2 marks)*

3 A student suggested that clones produced from somatic cell nuclear transfer would contain some DNA from the egg donor. Is the student correct? *(2 marks)*

22.4 Microorganisms and biotechnology
22.5 Microorganisms, medicines, and bioremediation

Specification reference: 6.2.1(e) and (f)

Scientists can exploit the reactions carried out by microorganisms for a variety of commercial and industrial processes.

Uses of microorganisms

▼ **Table 1** *Some of the uses of microorganisms in biotechnology*

Process	Examples	Microorganism used
Food production	Brewing (**alcohol** production) – anaerobic respiration by the fungus produces ethanol.	Brewer's yeast (a fungus)
	Baking (**bread** production) – the CO_2 produced by the fungus makes bread rise.	Baker's yeast (a fungus)
	Cheese and yoghurt production.	*Lactobacillus* bacteria
	Mycoprotein production (which can be eaten as a meat substitute).	*Fusarium* fungus
	Fruit juice – pectinase breaks down pectin in fruit and releases juice.	Pectinase enzyme from *A. niger* fungus
Drug manufacture	Penicillin **antibiotic**. (Antibiotics are secondary metabolites, which you will learn about in Topic 22.6, Culturing microorganisms in the laboratory.)	*Penicillium* fungus
	Insulin.	Genetically modifed *E. coli* bacteria
Bioremediation	**Water treatment**.	Various bacteria and fungi

Which characteristics make microorganisms useful?

▼ **Table 2** *The characteristics that make microorganisms useful in biotechnology*

Benefit of using microorganisms	Detail
Fewer ethical issues	The welfare issues associated with the use of animals are not present.
Short life cycle	Microorganisms reproduce rapidly; large numbers can be produced in a short period of time.
Genetic engineering	Bacteria are relatively easy to genetically engineer.
Simple nutrition	Nutrient requirements are cheap; they can often be grown on waste materials.
Fewer energy requirements	Most microorganisms require only low temperatures.

Summary questions

1 *Aspergillus* fungi can ferment soya beans, and this process is used to produce soya sauce. Explain why this is an example of biotechnology.
(1 mark)

2 Suggest why microorganisms are used to treat waste water (i.e. bioremediation).
(2 marks)

3 Suggest what advantage *Penicillium* may gain by producing antibiotic chemicals such as penicillin.
(2 marks)

22.6 Culturing microorganisms in the laboratory
22.7 Culturing microorganisms on an industrial scale

Specification reference: 6.2.1(g) and (h)

Scientists grow populations of microorganisms in the laboratory or on an industrial scale by culturing them (i.e. providing a nutrient medium that encourages their growth and replication).

Bacterial colony growth curve

The growth of bacterial populations follows a standard pattern and can be divided into four phases.

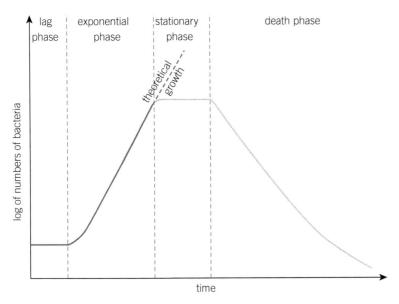

▲ **Figure 1** *A typical growth curve of bacterial populations*

▼ **Table 1** *The phases of a growth curve*

Phase	What happens?
Lag	Genes for important enzymes are transcribed, and bacteria adapt to their new environment.
Exponential (log)	The rate of reproduction is close to maximum and the population size increases at an exponential rate.
Stationary	The population reaches its maximum size (note: this could be referred to as *carrying capacity* in a natural environment); death rate = reproductive rate.
Death (decline)	Nutrients are exhausted, waste products are produced, and death rate rises.

Several **limiting factors** stop exponential growth, including: nutrient levels, oxygen availability, temperature, waste products, and pH.

Metabolite production

Metabolites are products of an organism's metabolism. **Primary metabolites** are substances formed as part of the normal growth of microorganisms (e.g. proteins, enzymes, and ethanol). Their rate of production follows the standard growth curve of bacteria. **Secondary metabolites** (e.g. antibiotic chemicals) are produced principally in the stationary phase. Not all microorganisms produce secondary metabolites.

> **Synoptic link**
>
> You have encountered the idea of limiting factors in Topic 17.4, Factors affecting photosynthesis, and you will read about them again in Topic 24.1, Population size.

Summary questions

1   Describe three factors that may result in the initiation of a death phase in a bacterial population. *(3 marks)*

2   Explain why the conditions in a *Penicillium* batch culture encourage the production of penicillin. *(3 marks)*

3   Bacterial cells were cultured in a 20 cm³ nutrient broth. 1 cm³ of the broth was placed in 9 cm³ of water to form a new solution (B). 1 cm³ of solution B was transferred to 9 cm³ of water to form solution C. 0.1 cm³ of solution C was placed on one agar plate and the same volume was transferred to another agar plate. Sixteen colonies grow on one plate, and twelve grew on the other plate. Estimate the population size in the original nutrient broth. Express your answer in standard form. *(4 marks)*

Laboratory cultures

The nutrient medium in which microorganisms are cultured can be liquid (broth) or solid (agar). Cultures need to contain the correct nutrients, be kept at the correct temperature, and be sterile. **Aseptic techniques** are used to keep the nutrient media sterile (e.g. preventing contamination of cultures from the air; using a sterilised inoculating loop to transfer bacteria to agar).

Worked example: serial dilutions

The number of cells in a population of microorganisms can be estimated by using serial dilutions. For example, imagine that we have a culture of bacteria in 10 cm³ of nutrient broth.

*   1 cm³ of the broth is transferred to 9 cm³ of water. This is a 10^{-1} dilution, and the new solution should contain 10% of the original cells.

*   1 cm³ of the new solution is transferred to 9 cm³ of water. This third solution will now contain 1% of the original cells.

*   0.1 cm³ of the final solution is then transferred to an agar plate. The agar plate should contain 0.01% of the original cells.
    10 cm³ / 0.1 cm³ = 100, and 1% / 100 = 0.01%.

*   40 colonies develop on the agar plate. We can assume each colony has grown from a single bacterial cell.

*   The number of cells in the original broth = (100% / 0.01% = 10 000) × 40 = 400 000.

Industrial cultures

On an industrial scale, microorganisms are cultured in reaction vessels called bioreactors.

▼ **Table 2** *Which factors are controlled in industrial cultures?*

What aspect is controlled?	Why is this controlled?
Temperature	To maintain the optimum temperature for enzymes.
Nutrients and oxygen	The medium is mixed and stirred to ensure an even distribution of respiratory substrates and oxygen. Levels can be monitored and extra nutrients and oxygen are added if necessary.
Asepsis	To prevent contamination from other, unwanted microorganisms.

Industrial cultures can be run using either **batch** or **continuous** culture.

▼ **Table 3** *A comparison of batch and continuous cultures*

	Batch	Continuous
How much nutrient medium is used?	A fixed volume of nutrient medium is used	Nutrient medium is added during the process
How long is the culture in the bioreactor?	A fixed time period	Indefinitely
What happens to the population size?	Waste and population builds up (but the process is halted before the death phase)	Microorganism population and waste are removed continuously to maintain population size
What can be produced?	Secondary metabolites (e.g. antibiotics)	Primary metabolites (and GM products such as insulin)

22.8 Using immobilised enzymes

Specification reference: 6.2.1(i)

As you learned in Topic 22.4, Microorganisms and biotechnology, enzymes from other organisms have been used by humans for many years. Rather than using the whole organism, enzymes can be isolated; this is more efficient. However, to improve efficiency even further, enzymes tend to be immobilised rather than free in solution.

Synoptic link

You learned about enzymes in Chapter 4, Enzymes. Types of bonding (e.g. ionic and covalent) were discussed in Topic 3.6, Structure of proteins.

Using immobilised enzymes

▼ **Table 1** *Advantages and disadvantages of using immobilised enzymes*

Advantages of using immobilised enzymes	Disadvantages of using immobilised enzymes
Less downstream processing (i.e. once the reaction has occurred, enzymes do not need to be separated from the product). Enzymes can be immediately **reused** (saving money). Enzymes can be more protected and therefore **more stable** and reliable.	**More expensive** and time-consuming to **set up**. Sometimes **less active** and therefore less efficient than freely dissolved enzymes.

How can enzymes be immobilised?

▼ **Table 2** *Methods of immobilising enzymes*

Method		What is done?	Pros	Cons
Surface immobilisation	Adsorption	Enzymes are attached to inert material (e.g. glass or alginate beads).	Relatively cheap.	Weak attachment increases the risk of enzymes leaking. Relatively slow.
	Covalent and ionic bonding	Enzymes form covalent bonds with silica gel or clay particles, or cross-linked to each other. Ionic bonds can form with cellulose or synthetic polymers.	Little enzyme leakage. The enzymes' active sites are very accessible for substrates.	Cost varies.
Entrapment	In matrix	Enzymes are trapped in a gelatin or cellulose matrix.	Active sites are not altered by the immobilisation.	Expensive. Active sites are less accessible for substrates. Diffusion and collection of the product can be slow.
	Membrane entrapment	Separation from the substrate solution using a semi-permeable membrane.		

Examples of immobilised enzyme use

Immobilised enzymes are especially useful for generating large quantities of product because continuous production is possible.

▼ **Table 3** *Examples of immobilised enzymes being used in biotechnology*

Enzyme	Product
Glucoamylase	Glucose (formed by breaking down dextrins)
Lactase	Glucose and galactose; lactose is broken down to produce lactose-free milk
Aminoacylase	L-amino acids (for drugs and cosmetics)
Penicillin acylase	Semi-synthetic penicillin (antibiotic)
Glucose isomerase	Fructose (which is used as a sweetener in food and drink)

Revision tip: Extra special

Extracellular enzymes are more commonly used than intracellular enzymes in research and for commercial processes. This is because extracellular enzymes are easier to isolate and are usually adapted to withstand a greater range of temperatures and pH than intracellular enzymes.

Summary questions

1   Describe how enzymes can be immobilised by entrapment. *(2 marks)*

2   Evaluate the economic advantages and disadvantages of using immobilised enzymes. *(3 marks)*

3   Immobilised penicillin acylase is used to produce semi-synthetic penicillin. Suggest why this may be very useful for the treatment of bacterial diseases. *(2 marks)*

1 Complete the following passage about natural cloning in plants by selecting the most appropriate words or phrases to place in the gaps.

(4 marks)

Many plants can produce clones naturally in a process known as vegetative .... The clones develop from undifferentiated tissue called .... For example, elm trees produce .... from their roots, which develop into .... identical clones of the original tree.

2 The following events form part of the somatic cell nuclear transfer procedure for producing artificial animal clones. Place the events in the correct chronological order. *(4 marks)*

   A The embryo is placed in a surrogate mother.

   B A nucleus is removed from one of the donor cells.

   C Electrofusion.

   D A cell is taken from an adult donor and cultured for several days.

3 Explain the changes in glucose and penicillin concentration shown in the graph.

(4 marks)

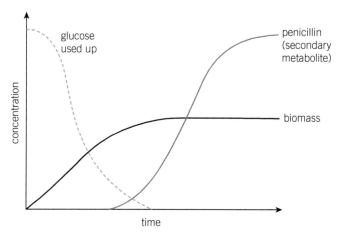

4 Suggest the form of enzyme immobilisation represented by each of the four diagrams below. *(4 marks)*

 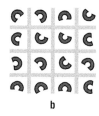 

a   b   c   d

23.1 Ecosystems
23.2 Biomass transfer through an ecosystem

Specification reference: 6.3.1 (a) and (b)

An ecosystem consists of organisms and the non-living (abiotic) factors with which they interact. As organisms consume other organisms, chemical energy (stored as biomass) is passed through an ecosystem. These energy transfers, however, are far from 100% efficient.

Factors affecting ecosystems

▼ **Table 1** *Examples of biotic and abiotic factors*

Factor	Examples	How does it affect organisms?
Biotic	Living organisms	Competition and consumption
Abiotic	Temperature	Enzyme activity Thermoregulation Leaf-fall and flowering (in plants)
	Light	Photosynthetic rate (in plants)

Food chains and trophic levels

A food chain illustrates the transfers of energy between organisms within an ecosystem. Each stage in a food chain is called a **trophic level**. Food chains vary in length. Two examples are shown below.

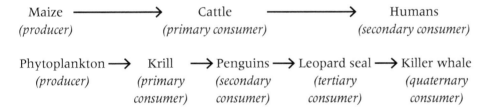

Maize ⟶ Cattle ⟶ Humans
(producer)   *(primary consumer)*   *(secondary consumer)*

Phytoplankton ⟶ Krill ⟶ Penguins ⟶ Leopard seal ⟶ Killer whale
(producer)   *(primary consumer)*   *(secondary consumer)*   *(tertiary consumer)*   *(quaternary consumer)*

Measuring biomass transfers between trophic levels

Transfers between trophic levels can be measured as biomass or energy (because biomass consists of stored chemical energy).

Biomass can be calculated as either wet or dry mass.

▼ **Table 2** *How biomass can be measured*

Measure of biomass	Procedure	Advantage	Disadvantage
Fresh (wet) mass	Living organisms measured.	No organisms are killed.	The presence of water reduces accuracy.
Dry mass	Organisms are killed and heated to 80°C until all water has been removed.	The estimates of mass are more accurate.	Organisms must be removed from the ecosystem and killed.

The trophic levels in a food chain are shown graphically in **pyramids**. The steps of the pyramid can represent numbers, biomass, or energy.

Key terms

Ecosystem: All of the biotic and abiotic factors in an area and their interactions.

Trophic level: A stage a food chain.

Producer: An organism (e.g. plant) that converts light energy to chemical energy (i.e. autotrophic nutrition).

Consumer: An organism that gains energy by feeding on other organisms (i.e. heterotrophic nutrition).

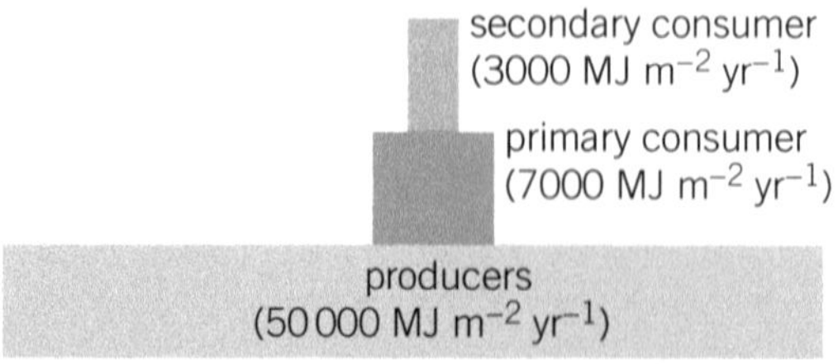

secondary consumer
(3000 MJ m^{-2} yr^{-1})

primary consumer
(7000 MJ m^{-2} yr^{-1})

producers
(50 000 MJ m^{-2} yr^{-1})

▲ **Figure 1** *A pyramid of energy*

Revision tip: When is a pyramid not a pyramid?

Pyramids of energy and biomass, if measured over a full year, will always be pyramid-shaped. Energy and biomass values decrease from producers up through the consumer levels. However, sometimes a pyramid of numbers will have a relatively narrow bar at the producer level. For example, trees may be few in number within an ecosystem, but their total biomass will still be greater than that of the primary consumers.

Maths skill: Units for energy and biomass

Biomass is usually measured in **g m^{-2} yr^{-1}** (grams per square metre per year) on land or **g m^{-3} yr^{-1}** (grams per cubic metre per year) in water.

Energy is usually measured in **kJ m^{-2} yr^{-1}** (or kJ m^{-3} yr^{-1}).

The units include a standard area (m^2) to enable comparisons to be made between ecosystems of different sizes.

You may be asked to convert between units. For example, 300 000 kJ m^{-2} yr^{-1} = 300 MJ m^{-2} yr^{-1} (because 1 MJ = 1000 kJ) = 300 000 000 MJ km^{-2} yr^{-1} (because there are one million square metres in one square kilometre) = 3 × 10^8 MJ km^{-2} yr^{-1} (in standard form).

Efficiency of transfers

Not all the energy in one trophic level is transferred to the next level. Some energy is released as heat, and some stored chemical energy (in biomass) cannot be consumed. The percentage of energy that is transferred represents the efficiency of the transfer.

▼ **Table 3** *The efficiency of energy transfers to producers and consumers*

	Efficiency at producer level	Efficiency at consumer levels
What transfer occurs?	Light energy is converted to chemical energy in producers.	The stored energy in the biomass of one trophic level is transferred to the next trophic level by consumption.
Why is some energy not transferred?	Most light energy (90%) cannot be absorbed by plants. Some absorbed energy is used in the reactions of respiration and is not converted to chemical energy in biomass.	Some parts of organisms (e.g. bones, roots, feathers) are inedible or indigestible. Some energy is lost from the food chain as heat, through movement, or in urine.
Calculation formula	Net production = gross production − respiratory losses	Ecological efficiency = (energy available after the transfer / energy available before the transfer) × 100

Worked example: Efficiency calculations

The net production in a grassland ecosystem was measured as $45 \, \text{g m}^{-2} \, \text{year}^{-1}$. In one year, a buffalo consumes the equivalent of all the biomass in a 350 m × 350 m area. 600 kg of grass is converted into biomass in the buffalo.

Calculate the efficiency of the energy transfer between the grass and the buffalo.

Step 1: Calculate the total biomass consumed by the buffalo.
$$45 \times (350 \times 350) = 5\,512\,500 \, \text{g} = 5512.5 \, \text{kg}.$$

Step 2: Calculate the efficiency of the transfer. $(600 \, \text{kg} / 5512.5 \, \text{kg}) \times 100 = 10.9\%$

How can humans manipulate biomass transfers?

Humans can tailor their farming practices to maximise production in plants and the efficiency of energy transfer to primary consumers.

▼ **Table 4** *How farming practices improve the efficiency of energy transfers*

	Which factor is manipulated?	How can humans manipulate the factor?
Production in plants	Light	Plants are grown in greenhouses under optimal light intensity and duration. Seed sowing is timed to maximise the leaf area present for photosynthesis during the brightest months of the year.
	Temperature	Greenhouses provide regulated, optimal temperatures.
	Water	Irrigation. Genetically engineered drought resistant crops.
	Nutrient levels	Fertiliser use.
	Pests	Pesticide use.
Efficiency of energy transfers to primary consumers	Movement	The movement of farm animals is limited. More energy is channelled into growth.
	Disease	Antibiotic use reduces energy expenditure in immune systems.

Summary questions

1 Describe the factors that reduce the efficiency of energy transfer between trophic levels. *(3 marks)*

2 Explain how farmers can increase the proportion of consumed energy that is used for growth in cattle. *(3 marks)*

3 Over a 2 year period, a farmer grew 200 000 kg of a crop in a 1 km² field.
   a Calculate the net primary production of this crop $(\text{kg m}^{-2} \, \text{yr}^{-1})$. *(2 marks)*
   b Explain how genetic engineering could increase the primary production of the crop. *(3 marks)*

Energy flows through an ecosystem and is transferred into the atmosphere. Nitrogen and carbon, however, are recycled back into ecosystems from the atmosphere.

Decomposition

Decomposition is the process of large organic molecules (in dead animals and plants) being broken down into smaller inorganic molecules. Decomposers can be either **bacteria** or **fungi**; they are also referred to as **saprotrophs** (i.e. they feed on dead or waste organic matter).

Decomposition in the nitrogen cycle is called **ammonification** (the breakdown of proteins, nucleic acids, and vitamins in dead organisms, faeces, and urine to form ammonia (NH_3) and ammonium (NH_4^+) compounds).

The nitrogen cycle

Nitrogen passes through a food chain as consumers feed on organisms in lower trophic levels. Decomposition temporarily removes nitrogen from a food chain. Three processes (outlined in the table) recycle nitrogen so that it eventually re-enters the food chain.

▼ **Table 1** *Reactions in the nitrogen cycle*

Process	What is the reaction?	Which bacteria carry out the reaction?
Nitrification	NH_3/NH_4^+ ions to NO_2^- (**nitrite** ions)	*Nitrosomonas*
	NO_2^- (nitrite ions) to NO_3^- (**nitrate** ions)	*Nitrobacter*
Nitrogen fixation	N_2 (in the atmosphere) is converted to NH_4^+ ions/NH_3	*Rhizobium* (mutualistic – lives in the roots of some plants)
		Azotobacter (free-living, in the soil)
Denitrification	NO_3^- to N_2 gas	Denitrifying bacteria (which require anaerobic conditions)

Revision tip: Oxidation or reduction?

You can also think about the processes in the nitrogen cycle in terms of chemistry – i.e. is nitrogen being oxidised or reduced? **Nitrification** represents **oxidation. Nitrogen fixation** and **denitrification** are examples of **reduction** reactions.

Revision tip: An absorbing choice?

Many plants tend to absorb nitrates (NO_3^- ions) from the soil. However, some plants can absorb ammonia (NH_3) /ammonium ions (NH_4^+) instead. This is especially true in soils with anaerobic conditions, where nitrification is difficult.

Remember, NH_3 is produced inside plant roots in species that contain *Rhizobium*, so it would be incorrect to write that these plants absorb NH_3 from soils.

Revision tip: Nitrogen and its compounds

It is easy to muddle the terms you will be using when describing the nitrogen cycle. Remember that 'nitrogen' is the element N, which is found in the atmosphere as a gaseous molecule, N_2.

Compounds containing nitrate ions (NO_3^-) or ammonium ions (NH_4^+) can be described as 'nitrogen-containing compounds' (e.g. potassium nitrate (KNO_3) and ammonium sulfate (NH_4SO_4)), as can ammonia (NH_3). If you are in doubt about any chemical formula you should use the full name.

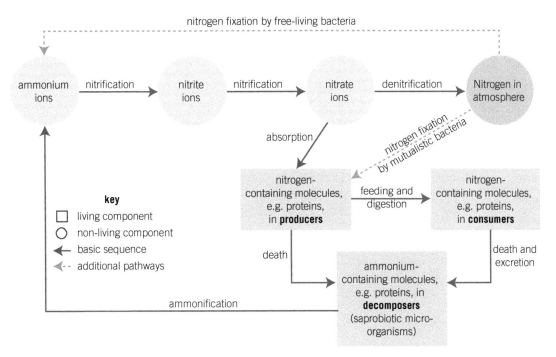

▲ **Figure 1** *The nitrogen cycle*

The carbon cycle

Photosynthesis converts CO_2 to organic molecules in organisms. **Respiration**, **decomposition**, and **combustion** return CO_2 to the atmosphere. These processes are naturally in balance, which maintains relatively constant atmospheric CO_2 concentrations. However, human activity (e.g. more combustion and deforestation) has increased the rate of CO_2 production.

Summary questions

1   Add the most appropriate words to complete this paragraph, which describes the nitrogen cycle.
    Ammonium ions can be added to the soil through two processes, _____ (controlled by bacterial species such as *Azotobacter*) and _____ (controlled by saprotrophic bacteria). Nitrifying bacteria convert ammonium ions to _____ ions and nitrate ions. In anaerobic conditions, nitrate ions are converted to nitrogen gas in a process known as _____ .   *(4 marks)*

2   Explain how human activity has upset the natural balance of the carbon cycle.   *(3 marks)*

3   Suggest why farmers may try to prevent their soils becoming waterlogged.   *(3 marks)*

The composition of an ecosystem gradually alters as the abiotic factors in the environment change. The progression from bare ground to a stable, complex community of organisms is known as **succession**.

Primary and secondary succession

Primary succession begins with bare ground (lacking soil) that has been newly formed (e.g. bare volcanic rock that becomes exposed). The starting point for secondary succession is bare soil that has resulted from deforestation or fire.

Question and model answer: Primary succession

Q. Describe the process of primary succession from bare rock to a stable community.

A.

- Only **pioneer species** can **colonise** the bare rock. These are species that are adapted to extreme conditions (e.g. exposure to wind and high light intensity) and can fix nitrogen from the atmosphere. Examples include lichen and algae. Pioneer species represent the first **seral stage**.

- The erosion of the rock produces a basic soil.

- The death and decomposition of pioneer organisms adds nutrients to the soil.

- Soil development enables other species to colonise (i.e. secondary colonisers such as mosses).

- Improved environmental conditions enable more species to colonise (which outcompete the pioneers and secondary colonisers).

- Eventually a **climax community** forms. This is stable (i.e. if environmental conditions remain the same, the species numbers will not vary much), and relatively biodiverse.

Deflected succession

Human activities can prevent a climax community from forming. This is known as **deflected succession** and results in the formation of a community known as a **plagioclimax**. Two examples of plagioclimax communities are found on agricultural land and in managed forests.

Revision tip: What about animals?

Succession also occurs among animal communities. As the composition of plant species in a community develops, this enables different animal species to enter the ecosystem and exploit new niches.

Summary questions

1 Explain what is meant by a plagioclimax, and identify one example resulting from land management. (*2 marks*)

2 Describe three important characteristics of a pioneer species. (*3 marks*)

3 Describe the difference between primary succession and deflected succession, and explain why primary succession is likely to result in greater biodiversity. (*3 marks*)

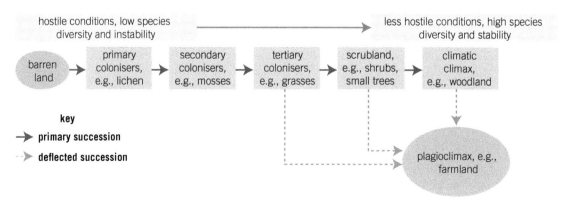

▲ **Figure 1** *Ecological succession*

23.5 Measuring the distribution and abundance of organisms

Specification reference: 6.3.1 (e)

You learned about sampling techniques and strategies in Chapter 11, Biodiversity. Here you will examine how these methods are employed to measure the distribution and abundance of organisms.

Measuring distribution

A **transect** is used to analyse how the distribution of organisms varies in an ecosystem. This is a form of **systematic** sampling (see Topic 11.2, Types of sampling).

Measuring abundance

Sampling allows population sizes to be estimated. **Quadrats** can be used to estimate the abundance of a plant species per area (see Topic 11.3, Sampling techniques).

 Worked example: Estimating plant abundance

A 1 m² quadrat was used to take 8 samples in a 200 m² field. 70 dandelion plants were counted. Estimate the dandelion population size in the field.

Step 1: Individuals m⁻²
= Number of individuals counted / total area sampled (m²) = 70/8 = 8.75 m⁻²

Step 2: Multiply the calculated value by the total area of the habitat. i.e.
8.75 × 200 = 1750

Animal abundance is estimated using the **capture-mark-release-recapture** technique.

 Worked example: Estimating animal abundance

Scientists captured and marked 13 water voles along a river ecosystem. One week later they captured 14, of which 2 were marked.

Estimate the total vole population in the river.

Step 1: Choose the correct equation, which is:

Number of individuals in the first sample × number in the second sample / number of marked individuals in the second sample.

Step 2: Enter the measured values into the equation: 13 × 14 / 2 = 91

The total vole population is estimated to be 91.

Summary questions

1 A 1 m² quadrat was used to take 500 samples in a 1 km² grassland. 48 blue fleabane plants were counted. Estimate the blue fleabane population size in the grassland. *(2 marks)*

2 Explain why the measurement of distribution should be systematic. *(2 marks)*

3 A student captured and marked 8 slugs in a field. One week later the student captured 16, of which 3 were marked. Estimate the total slug population in the field. *(2 marks)*

Revision tip: Which transect?
Two forms of transect exist. A **line transect** involves taking samples at regular intervals along a linear tape. It will generally show distribution but give no information about abundance. A **belt transect** involves recording species within two parallel lines. It provides both distribution and abundance data.

Revision tip: Calculating biodiversity
Once abundance data have been collected for all the species in a habitat, biodiversity can be calculated using Simpson's Index of Diversity (see Topic 11.4, Calculating biodiversity).

Synoptic link
You learned the fundamentals of ecological sampling in Topics 11.2, Types of sampling, and 11.3, Sampling techniques. Topic 11.4, Calculating biodiversity, covered how to use the Simpson's Index of Diversity equation.

1   A farmer grew 80 000 kg of a crop in a 480 m² field over a period of one year. What is the net primary production of this crop (units = g m⁻² yr⁻¹)?   (*1 mark*)

   A 1.67×10^2 **C** 1.67×10^4

   B 1.67×10^3 **D** 1.67×10^5

2

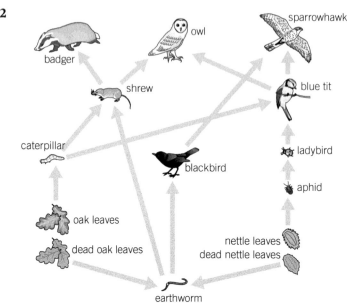

The diagram shows part of a woodland food web.

   a   Name all of the producers shown in the food web.   (*1 mark*)

   b   Name all of the tertiary consumers shown in the food web.   (*1 mark*)

   c   Suggest the short-term effects on the populations of other species if the ladybird population were to disappear from this ecosystem.   (*4 marks*)

   d   Suggest which energy transfer in this food web is likely to be the least efficient. Explain your answer.   (*2 marks*)

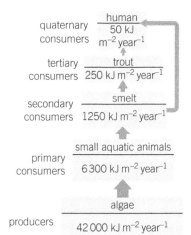

3   Use to the diagram on the left to calculate the ecological efficiency of energy transfer between:

   a   smelt and humans   (*1 mark*)

   b   trout and humans   (*1 mark*)

   c   algae and small aquatic animals.   (*1 mark*)

4   Match the following sentences to the correct process in the nitrogen cycle.   (*3 marks*)

   A   Carried out by *Nitrobacter*.

   B   Carried out by *Nitrosomonas*.

   C   Produces nitrogen gas.

   D   Produces nitrate ions.

   E   Produces ammonia.

   F   Removes nitrogen gas.

   G   Removes nitrate ions.

   Processes:

   Nitrification   Denitrification   Nitrogen fixation

24.1 Population size

Specification reference: 6.3.2 (a)

The size of a population is dictated by many biotic and abiotic factors, some of which limit the number of individuals the population can support. The maximum size of a population is known as its **carrying capacity**.

Population growth curves

Growth curves tend to be sigmoidal in shape. At first, populations grow slowly before experiencing a period of rapid growth. Eventually, population size plateaus when the carrying capacity is reached; numbers will fluctuate during this stable state but remain relatively constant.

Factors that determine population size

Limiting factors determine the size to which a population can grow.

▼ **Table 1** *Density-dependent and density-independent factors*

	Definition	Examples
Density-dependent factors	The impacts of these factors vary with population density.	**Abiotic**: temperature, water availability, pH, light intensity. **Biotic**: predation, disease, competition, migration.
Density-independent factors	Factors that affect a population regardless of its size.	Natural events (e.g. earthquakes, storms).

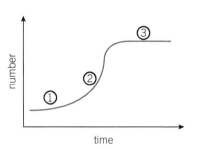

▲ **Figure 1** *A typical population growth curve. 1 = slow growth (lag phase). 2 = rapid growth. 3 = a stable state (carrying capacity); birth rate and death rate are very similar*

Synoptic link

Animal population growth curves resemble the growth curves of bacterial colonies, which you examined in Topic 22.6, Culturing microorganisms in the laboratory.

Key term

Limiting factor (in the context of populations): A factor that restricts the final size of a population.

Summary questions

1 Define **a** a limiting factor **b** a density-dependent factor. *(2 marks)*

2 Examine the demographic graph in Figure 2. Describe and explain the change in population size during stages 2 and 3. *(3 marks)*

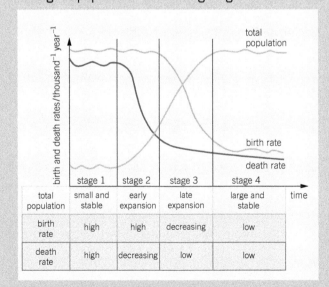

▲ **Figure 2**

3 Population pyramids illustrate the distribution of ages within populations. The two graphs in Figure 3 show age distribution in two countries with similar populations. State which country has the faster population growth. Explain your answer. *(2 marks)*

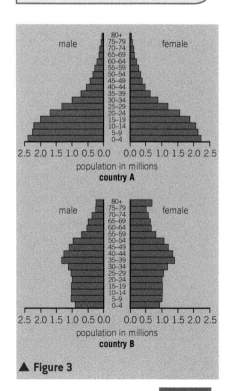

▲ **Figure 3**

24.2 Competition
24.3 Predator–prey relationships
Specification reference: 6.3.2 (b)

Members of a population compete with each other for limiting factors such as food, light, and space. Populations also interact with other species by competing for resources and through predator–prey relationships.

Intraspecific and interspecific competition

Interspecific competition = **different species** competing for resources. The competing species overlap in their food sources, behaviour, or the habitat they occupy. The better adapted species may **outcompete** the other species if resources are limited.

Intraspecific competition = members of the **same species** competing for limited resources.

Revision tip: Interspecific vs. intraspecific

It is important to specify the type of competition that is occurring. Remember that 'interspecific' means 'between species' (think of 'intervals' or 'inter-city trains' perhaps), and 'intraspecific' means 'within species'.

Go further: Allelopathy – competition between plants

Some plant species enhance their ability to compete with other plants by releasing toxic or inhibitory chemicals into their surroundings, which is called **allelopathy**. Examples of this chemical warfare include the release of substances that inhibit chlorophyll production and molecules that reduce the rate of respiration. The black walnut, *Juglans nigra*, has several allelopathic tools at its disposal. It releases a chemical called juglone into the soil, which inhibits respiration in nearby plants, and also contains allelopathic chemicals in its leaves and roots.

1 Suggest how the storage of toxic chemicals in leaves can result in allelopathy.

2 Suggest why the species growing in a habitat should be assessed for their allelopathic properties before the land is selected for agriculture.

Summary questions

1 State three factors for which plant species are likely to compete. (*3 marks*)

2 Corals, limpets, and anemones are all filter feeders (they strain food particles from water) with habitats on the ocean floor. Barnacles are filter feeders that have evolved the ability to form colonies on the skin of whales and the side of ships. Explain the benefit to barnacles of this ability. (*3 marks*)

3 Analyse Figure 1 and explain why the peaks in predator numbers are lower than those of prey, and explain why the peaks are delayed compared to the peaks in prey numbers. (*3 marks*)

Analysing predator–prey relationships

All predator–prey relationships show a similar pattern: numbers in both populations fluctuate, but the peaks and troughs in the predator population are delayed in comparison to those of the prey.

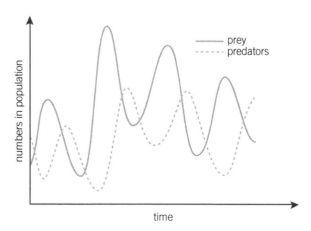

▲ **Figure 1** *A typical predator–prey graph*

24.4 Conservation and preservation
24.5 Sustainability

Specification reference: 6.3.2 (c) and (d)

Ecosystems can be protected through **conservation** or **preservation**. These two terms are often used interchangeably, but they have distinct meanings.

Conservation or preservation?

▼ **Table 1** *A comparison of conservation and preservation*

	Definition	What is done?
Conservation	The sustainable management of ecosystems to maintain biodiversity.	Active, **sustainable** management (i.e. balancing the maintenance of biodiversity with the extraction of resources). This can include **reclamation** (i.e. restoring damaged ecosystems).
Preservation	The maintenance of ecosystems in their original state (without interference).	Ecosystems are monitored, but visitors are not allowed, and interference is kept to a minimum.

The importance of conservation

Conservation is important for a range of ethical, social, and economic reasons.

Sustainability

Ecological sustainability is the exploitation of resources without compromising biodiversity or the ability to meet future requirements (i.e. ensuring that resources are renewable and will not run out).

▼ **Table 2** *Examples of sustainable practices*

	How is it done sustainably?
Timber production	**Coppicing** (for small-scale production), which involves trees being cut close to the ground. The trees remain alive and eventually produce new shoots. Coppicing can be **rotational** (i.e. the area being coppiced varies each year). Biodiversity remains high because succession is stopped. On a larger scale, trees are felled and will not regrow. To ensure sustainability, trees are **replanted** and only the largest trees are cut each year.
Fishing	Fishing **quotas** set limits on the number of fish of certain species that can be caught. Sustainability can also be improved through the use of **fish farms** (rather than catching wild fish) and promoting the consumption of species that are lower in the food chain.

Summary questions

1   Describe the difference between conservation and preservation.

(2 marks)

2   Tilapia fish are primary consumers. Salmon are tertiary consumers. Explain why the farming of tilapia is more sustainable than the farming of salmon.

(2 marks)

3   Suggest why reclamation of a habitat is difficult. *(2 marks)*

Question and model answer: The importance of conservation

Q. Describe the ethical, social, and economic benefits of conserving ecosystems.

A.

> Try to avoid vague answers such as "we should not play God" or "animals have the right to live".

Ethical – Humans have a **moral responsibility** to maintain biodiversity. (However, the ethical argument for conservation is subjective.)

Social – Conserved ecosystems provide natural beauty and are **aesthetically attractive**. They can provide amenities/**recreation**, enable **ecotourism**, and have an **educational** benefit.

Economic – Organisms can be harvested from ecosystems using a **sustainable** approach. Humans gain resources such as **food, drugs**, and timber. The maintenance of biodiversity increases the chance of discovering organisms with useful properties that have not yet been recognised.

Key terms

Conservation: Protection of an ecosystem through active management.

Preservation: Protection of an ecosystem by preventing human use or interference.

Sustainability: The use of a natural resource without damaging biodiversity and ensuring that the resource is not depleted.

Ecosystems often require management to balance the maintenance of biodiversity with human economic and social needs. Here you will learn about how this potential conflict is being dealt with in a range of different ecosystems.

Ecosystem management case studies

Ecosystem	Description of the ecosystem	Conservation methods
Masai Mara, Kenya	A savannah ecosystem, containing grassland and woodland. Large populations of zebra, buffalo, elephants, leopards, and lions.	Promotion of **ecotourism** (which attempts to balance conservation with tourism, as well as benefitting local people). Prevention of rhino **poaching** (e.g. by employing a nature reserve ranger). **Legal hunting** is allowed in some cases, which culls animals considered to be in excess.
Terai region, Nepal	High temperatures and humidity in the summer months. Fertile soils with lush forest. High biodiversity. Heavy deforestation and agricultural use.	**Sustainable forest management** (e.g. harvesting quotas). Improved **irrigation** schemes (to increase the efficiency of agricultural production). Encouragement of fruit and vegetable growing in nearby regions to **relieve the pressure of intensive farming** on the Terai region.
Peat bogs	Ground with high water content and decomposing vegetation. Natural stores of CO_2. Peat can be removed for use as a fuel to improve farm soils. Contains rare species.	**Blocking ditches** that have been constructed to prevent flooding on neighbouring land. This **raises water levels** in the peat bog. Removal of trees and **prevention of afforestation** (because trees have a high water demand and reduce the water content of bogs).
The Galapagos islands	An **environmentally sensitive ecosystem** (i.e. especially vulnerable to change). Contains unique species (e.g. the Galapagos giant tortoise).	Limiting and managing tourism, with park rangers to monitor conservation projects. Strict controls over the movement of introduced animals (e.g. goats and pigs).

Summary questions

1   Describe what is meant by an environmentally sensitive ecosystem.

(1 mark)

2   Local tribes in the Masai Mara have traditionally grazed livestock using semi-nomadic farming, which involves moving area as the climate changes. Nowadays, local tribes are restricted to the edges of the nature reserve. Suggest the impact this may have on the ecosystem.

(3 marks)

3   Suggest why both conservation and preservation of peat bogs is difficult to achieve.

(3 marks)

Revision tip: Applying conservation principles

Most principles of conservation and ecosystem management apply to all the case studies outlined here. You should be prepared to apply these ideas to other ecosystems, but try to learn the specific details of the examples on this page.

1 Which of the following statements represents a density-independent factor that can influence population size? (*1 mark*)

   A Competition

   B The spread of disease

   C Climate

   D Parasitism

2 Describe how the rise in human population has affected abiotic factors in ecosystems and explain the impact these changes have had on biotic factors. (*6 marks*)

3 The graph below illustrates the effects of intraspecific competition over time. Describe what is occurring at stages 1–3. (*3 marks*)

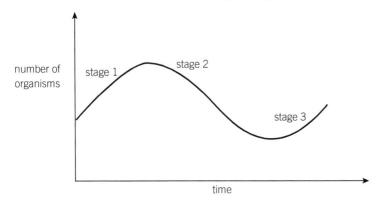

4 The graph shows a general predator–prey graph. Explain the changes to prey and predator populations at stages 1–4. (*4 marks*)

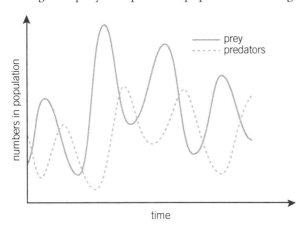

Answers to summary questions

13.1

Go further: Types of cell signalling

1 a Paracrine signalling; the presynaptic and postsynaptic neurones are close to each other, separated only by the synaptic cleft. Neurotransmitters diffuse between the two neurones.

  b Endocrine signalling; insulin (a hormone) is released from beta cells and passes through the blood to target cells in the liver.

Summary questions

1 Neurotransmitters; *[1]*

  Postsynaptic; *[1]*

  Endocrine *[1]*

2 Transmits nervous impulses between neurones; *[1]*

  Across synapses / synaptic clefts; *[1]*

  For coordinated responses to stimuli; *[1]*

  idea of summation *[1][2 max]*

3 Receptor cells need to communicate with controller / coordinator / brain; *[1]*

  Controller / coordinator / brain needs to communicate with effectors; *[1]*

  idea of several cell types / tissues / systems involved in responses *[1] [2 max]*

13.2

1 Position of cell body (described); *[1]*

  Sensory neurone has one (long) dendron; *[1]*

  Motor neurone has many dendrites leading to its cell body *[1] [2 max; reverse arguments apply]*

2 Relay neurones tend to be short; *[1]*

  The increased rate of conduction in the presence of myelin sheath would make little difference *[1]*

3 a Proteins are synthesised in the cell body; *[1]*

    Some proteins must travel to other regions of the neurone (e.g. motor junctions at the other end of the cell); *[1]*

  b Vesicle (production) / cytoskeleton / motor proteins *[1]*

13.3

1 (Sensory receptors) convert one form of energy to another form of energy *[1]*

2 a Light; *[1]*

  b Heat / thermal; *[1]*

  c Mechanical pressure; *[1]*

  d Chemical *[1]*

3 *In response to hormones*:

  cAMP is a second messenger; *[1]*

  stimulates cascade of reactions (in target cell) *[1]*

  In sensory receptor:

  Alters cell membrane permeability; *[1]*

  To sodium ions; *[1]*

  Stimulates depolarisation / action potentials *[1] [3 max]*

13.4

1 three; *[1]*

  two; *[1]*

  depolarised; *[1]*

  repolarisation *[1]*

2 *(with myelination)*

  Faster impulses; *[1]*

  Saltatory conduction; *[1]*

  Longer local circuits; *[1]*

  Nodes of Ranvier present
  [1] [3 max] (allow reverse argument throughout)

3 All-or-nothing principle; *[1]*

  Threshold potential must be surpassed; *[1]*

  No action potential will occur if Na^+ ion influx is insufficient; *[1]*

  Stimulus strength does not affect action potential magnitude (above the threshold); *[1]*

  K^+ ion channels always open at the same point in an action potential (which results in the potential difference never surpassing a particular value)
  [1] [3 max]

13.5

1 Calcium ion channels open / calcium ions diffuse into the presynaptic neurone; *[1]*

  Vesicles fuse with the presynaptic membrane; *[1]*

  Acetylcholine is released by exocytosis *[1]*

2 A weak stimulus produces less frequent action potentials; *[1]*

  Fewer neurotransmitter molecules are released; *[1]*

  Less depolarisation (in postsynaptic neurone); *[1]*

  Threshold is not reached; *[1]*

  No postsynaptic action potential *[1] [3 max]*

3 To be released again from the presynaptic neurone; *[1]*

  Regulates the concentration of neurotransmitter in the synaptic cleft; *[1]*

  Prevents re-binding to receptors when signalling has stopped *[1] [2 max]*

13.6/13.7

1 a somatic; [1]

b sympathetic; [1]

c parasympathetic [1]

2 The cerebrum controls voluntary movements; [1]

The cerebellum coordinates balance and non-voluntary movements [1]

3 Humans have a larger brain than expected for their body mass; [1]

This may represent an evolutionary adaptation for problem solving / higher thought processes / language / complex learning / social interactions [1]

13.8

1 a spinal cord; [1]

b brain stem [1]

2 Few synapses; [1]

Short pathway; [1]

A dangerous stimulus is responded to quickly; [1]

Innate /no learning required; [1]

The response is stereotyped/consistent/always the same; [1]

Involuntary; [1]

Prevents overloading of the brain [1] [max 4]

3 Two synapses; [1]

Short relay neurone; [1]

Sensory neurone does not need to travel deep into the brain; [1]

Response is quicker [1] [max 3]

13.9/13.10

1 a no change; [1]

b shortens; [1]

c no change; [1]

d no change; [1]

e shortens [1]

2 Calcium ions bind to troponin; [1]

Changes the shape of troponin; [1]

Displaces tropomyosin; [1]

Uncovers myosin binding sites on actin; [1]

Enables myosin heads to bind to actin [1] [max 3]

3 (Muscle fibres have)

More mitochondria; [1]

To generate ATP in aerobic respiration [1]

More ribosomes; [1]

For actin/myosin/troponin/tropomyosin production [1]

Sarcoplasmic reticulum; [1]

To store and release calcium ions [1]

Myofibrils; [1]

To act as contractile organelles [1]

Many nuclei; [1]

Due to the fusion of muscle cells
[1] [max 6] (accept reverse arguments when relevant)

14.1

1 a The cell to which a hormone binds and in which it produces an effect; [1]

b A substance that is activated by a hormone; [1]

And produces a response within a target cell [1]

2 *Similarities:*

transported in blood; [1]

produce responses in target cells; [1]

bind to receptors; [1]

regulate transcription [1] [max 2]

Differences:

steroid hormones diffuse through cell membranes (non-steroid hormones do not); [1]

steroid hormone receptors are in the cytoplasm or nucleus / non-steroid hormone receptors are on the cell surface membrane; [1]

non-steroid hormones activate second messengers / reaction cascades [1] [max 2]

3 Glucocorticoids secreted, which regulate carbohydrate metabolism; [1]

More (named) respiratory substrates made available; [1]

Mineralocorticoids (e.g. aldosterone) enable more water to be reabsorbed from the kidneys; [1]

Adrenaline can raise heart rate (during the race); [1]

Adrenaline promotes glycogenolysis / raises blood glucose concentration; [1]

Noradrenaline increases heart rate / widens airways [1] [max 5]

14.2

1 Differential staining; [1]

Islets of Langerhans stain blue/lilac and acini stain dark pink/purple; [1]

Acini cells are arranged in small clusters / cells in the islets are arranged in larger clusters [1] [max 2]

2 Endocrine glands secrete hormones into the blood; [1]

exocrine glands secrete other substances such as enzymes; *[1]*

the pancreas operates as both types of gland; *[1]*

it secretes digestive enzymes; *[1]*

it secretes the hormones insulin and glucagon *[1] [max 4]*

3 β cells have potassium ion channels that close when glucose concentrations are high; *[1]*

β cells have calcium ion channels that open when glucose concentrations are high; *[1]*

β cells transport insulin by exocytosis; *[1]*

β cells have glucose receptors/transporters; *[1]*

α cells transport glucagon by exocytosis; *[1]*

α cells (also) have glucose receptors/transporters; *[1]*

idea that α cells secrete glucagon when little glucose is entering the cells *[1] [max 5]*

14.3

1 (ATP is) produced when glucose enters the cell; *[1]*

Binds to potassium ion channels; *[1]*

Which causes depolarisation / opens calcium ion channels *[1] [max 2]*

2 9×10^{-4} (or 0.0009) *[1]*

3 Liver cells contain stores of glycogen; *[1]*

glucagon is secreted by the pancreas in response to low blood glucose concentration; *[1]*

when glucagon binds to receptors on liver cells, glycogen within these cells is broken down into glucose molecules *[1]*

14.4

Go further: Maturity-onset diabetes of the young

1 The condition is inherited via a single gene, whereas type 1 and type 2 diabetes appear to have more complicated genetic influences (e.g. it is likely that many different genetic variants increase a person's susceptibility for developing type 2 diabetes).

2 0.75–1.00 / 75–100% (0.75 if both parents are heterozygous and 1.00 if either parent is homozygous dominant for the MODY allele).

3 The homozygous genotype (i.e. having two copies of the defective MODY gene variant) produces symptoms that are more severe.

Summary questions

1 Type 1 is insulin-dependent because people with this form produce insufficient insulin; *[1]*

type 2 is insulin-independent because people with this form often produce sufficient insulin but the target cells are insensitive to the hormone *[1]*

2 Blood glucose increases to a greater extent in the diabetic; *[1]*

blood glucose takes longer to return to the original level in the diabetic; *[1]*

the diabetic produces no additional insulin because their pancreatic beta cells have been destroyed *[1]*

3 Type 1 diabetes is not affected by lifestyle to a great extent; *[1]*

the prevalence of type 1 diabetes is unlikely to change much; *[1]*

the prevalence of type 2 diabetes is increasing; *[1]*

this is because of an aging population and an increase in obesity levels; *[1]*

the increasing prevalence will cost health care services more money to treat and manage in the future. *[1]*

14.5

1 (Adenylyl cyclase is) activated by (the binding of) adrenaline; *[1]*

on liver cell surface membrane; *[1]*

(Adenylyl cyclase) converts ATP; *[1]*

into cAMP *[1] [max 2]*

2 Both systems work together to produce responses; *[1]*

The sympathetic nervous system stimulates the release of hormones *[1]*

3 Hearing is not (usually) essential for responding to threats; *[1]*

Blood flow is diverted from the auditory region of the brain to other regions *[1]*

14.6

1 Chemoreceptors detect changes in blood pH that reflect carbon dioxide levels; *[1]*

pressure receptors detect changes in blood pressure; *[1]*

the receptors pass information to the medulla oblongata, which initiates the necessary response *[1]*

2 Increased physical activity raises the amount of energy required by muscles; *[1]*

muscles require more oxygen and glucose for respiration; *[1]*

cardiac output must be increased to supply muscles with these molecules at the necessary rate *[1]*

3 Blood pressure would remain high; *[1]*

the parasympathetic nervous system is no longer able to carry impulses / transmit to the heart; *[1]*

the medulla oblongata cannot stimulate the SAN to lower heart rate *[1]*

15.1

1 a The desired value around which negative feedback operates. *[1]*

 b The range of values across which a physiological factor varies and negative feedback operates. *[1]*

2 Oxytocin increases the strength and frequency of contractions; *[1]*

 the increase in contractions causes more oxytocin to be released; *[1]*

 contractions are therefore intensified further and gradually increased; *[1]*

 the gradual increase in contractions enables a baby to be born using the minimum intensity of contraction. *[1]*

3 Homeostasis maintains conditions within a narrow range around an optimum value (set point); *[1]*

 negative feedback reverses any deviation away from the set point; *[1]*

 positive feedback increases any change made to a physiological factor. *[1]*

15.2/15.3

1 Arterioles (and shunt vessels) ; *[1]*

 Sweat glands ; *[1]*

 Hair erector muscles ; *[1]*

 Fat tissue *[1]*

2 *Advantages*

 Lower food requirements (than endotherms); *[1]*

 A greater proportion of energy intake can be used for growth *[1]*

 Disadvantages

 Lower activity levels in cold temperatures (compared to endotherms); *[1]*

 (Therefore) greater risk of predation; *[1]*

 May need to survive winter without food intake (due to lack of activity) *[1] [max 4]*

3 More heat is lost to the environment; *[1]*

 More food provides additional substrates for respiration; *[1]*

 More respiration increases metabolic rate; *[1]*

 More heat is generated (in response to the colder winter temperatures); *[1]*

 Some species, especially hibernating species, increase their fat reserves *[1] [max 3]*

15.4

1 Hydrogen peroxide; *[1]*

 ethanol; *[1]*

 a (named) drug *[1] [max 2]*

2 (Transamination is) the conversion of one amino acid into another; *[1]*

 some amino acids are not supplied in an organism's diet *[1]*

3 In hepatocytes / intracellular enzymes; *[1]*

 enzymes control deamination and the reactions of the ornithine cycle; *[1]*

 (glycogen synthase) catalyses glucose to glycogen conversion; *[1]*

 (glycogen phosphorylase) catalyses glycogen to glucose conversion; *[1]*

 Catalase; *[1]*

 breaks down hydrogen peroxide ; *[1]*

 Alcohol dehydrogenase ; *[1]*

 removes ethanol *[1] [max 5]*

15.5

Go further: Diuretics – treating, cheating, and energy-depleting?

1 Fewer sodium and chloride ions are co-transported from the distal convoluted tubule into the tissue fluid; the DCT water potential remains lower than it otherwise would be; less water diffuses out of the nephron by osmosis.

2 Alcohol inhibits ADH, which causes a greater volume of urine to be produced; drinking water replaces some of the lost fluid; this helps to lower the solute concentration of blood.

Summary questions

1 The renal vein will have a lower concentration of urea (and other toxins); *[1]*

 water potential may be different in the two blood vessels *[1]*

2 To transport (the few) proteins that have been filtered into the nephron; *[1]*

 endocytosis transports proteins from the PCT lumen into cells in its wall; *[1]*

 exocytosis transports proteins from cells in the PCT wall into tissue fluid *[1]*

3 Dialysis does not provide a cure, only an improvement in condition for the patient; *[1]*

 successful kidney transplantation provides a cure; *[1]*

 however, transplant surgery carries a high risk of organ rejection; *[1]*

 future transplant surgery may use therapeutic cloning to reduce the possibility of rejection; *[1]*

 haemodialysis usually requires regular trips to a health clinic and is time-consuming; *[1]*

 peritoneal dialysis can be carried out by the patient while they work, but must be done every day *[1]*

16.1

1 Cell elongation;   [1]

   apical dominance / shoot growth;   [1]

   root growth;   [1]

   (photo)tropism(s);   [1]

   ethene release / fruit ripening;   [1]

   prevention of abscission / leaf fall   *[1] [max 3]*

2 *Similarities [max 2]*

   Cell signalling molecules;   [1]

   receptors;   [1]

   idea of often work as antagonists/in opposition (e.g. glucagon and insulin, gibberellin and ABA) *[1]*

   Differences [max 2]

   Animals have endocrine glands / plants lack glands;   [1]

   animal hormones move in blood and plant hormones move in phloem/xylem/through cells; [1]

   animal hormones produce effects more rapidly   [1]

3 0.005 (5×10^{-3}) $mol\,dm^{-3}$ [1]

   $0.000\,05$ (5×10^{-5}) moles   [1]

16.2/16.3

Go further: Flowering – the long and the short of it

1 Long-day plants: The proportion of P_{fr} increases as day length increases and night length decreases, which initiates flowering.

   Short-day plants: The proportion of P_r increases as day length decreases and night length increases, which initiates flowering.

Summary questions

1 Physical defences;   [1]

   thorns/barbs/spikes/stings;   [1]

   fibrous/inedible tissue;   [1]

   chemical defences;   [1]

   (named) toxic compound;   [1]

   pheromones for communication between plants   *[1] [max 4]*

2 Auxin concentration reduced;   [1]

   increased ethene sensitivity (in abscission zone); [1]

   cellulase/digestive enzyme genes switched on/transcribed;   [1]

   increased cellulase production;   [1]

   cell walls digested in abscission zone   *[1][max 4]*

3 Pheromones enable communication between plants;   [1]

   neighbouring plant produces callose (before infection)   [1]

16.4

1 Plant is grown on its side;   [1]

   constant rotation;   [1]

   idea that gravitational force is felt evenly by all parts of the plant;   [1]

   roots and shoots grow straight / horizontally   *[1] [2 max]*

2 Transport of auxin from tip/apex down the shoot; [1]

   protons/H^+ ions pumped into cell walls;   [1]

   causing cell elongation   [1]

3 Positive phototropism increases photosynthetic rate;   [1]

   more carbohydrates produced;   [1]

   idea that negative phototropism causes roots to grow further into soil;   [1]

   more water/minerals absorbed   [1]

16.5

1 Promotes/speeds up fruit dropping at high concentrations;   [1]

   slows down/prevents fruit dropping at low concentrations;   [1]

   slows down leaf fall;   [1]

   promotes root growth;   [1]

   weedkillers   *[1] [3 max]*

2 Ethene cannot be sprayed in liquid form;   [1]

   gas release is less controlled   [1]

3 Increases/speeds up ripening [no mark]

   Explanation:

   Rise in carbon dioxide indicates an increase in respiration rate;   [1]

   which indicates polysaccharides are being broken down to glucose/sucrose/maltose/fructose/respiratory substrates (during the ripening process)   [1]

17.1

1 Endothermic;   [1]

   active;   [1]

   ATP   [1]

2 Use of electron transport chains;   [1]

   use of coenzymes (e.g. NAD in respiration and NADP in photosynthesis);   [1]

   some of the same intermediates are formed in both (e.g. GP);   [1]

   starting materials are regenerated in both;   [1]

   both use chemiosmosis;   [1]

   to produce ATP   *[1] [max 3]*

3 Light energy (photons) converted to chemical energy in organic molecules in photosynthesis; *[1]*

respiration converts chemical energy in organic molecules into chemical energy in inorganic molecules (ATP); *[1]*

movement of H^+ ions (kinetic energy) drives ATP production via chemiosmosis; *[1]*

chemical energy is converted to thermal energy via metabolic reactions *[1] [max 3]*

17.2

1 Excited electrons have different origins (i.e. light absorption in photosynthesis, and from organic reactions in respiration); *[1]*

locations differ (i.e. thylakoid membranes in photosynthesis, and inner mitochondrial membranes in respiration) *[1]*

2 Excited electrons are passed along electron transport chains; *[1]*

energy is released as the electrons are passed to lower energy levels; *[1]*

the energy is used to pump protons; *[1]*

from the stroma to the thylakoid space/lumen *[1] [max 3]*

3 a The intermembrane space has a greater concentration of H^+ ions; *[1]*

this suggests H^+ ions have been transported into the space from the matrix *[1]*

b The positive charge of the protons in the intermembrane space creates the potential difference *[1]*

17.3

Go further: A visit to the GP

1 a ATP

b NADPH (reduced NADP)

2 CH_2OH

3 NO_3^- (nitrate ions) would be required to provide N for the amine group

Summary questions

1 Glycerol; *[1]*

triglyceride/phospholipid/plasma membrane production *[1]*

Glucose/fructose; *[1]*

respiratory substrates/polysaccharide formation *[1]*

Amino acids; *[1]*

protein/enzyme formation *[1] [max 4]*

2 A = 0.750 *[1]*

B = 0.375 *[1]*

3 ATP and NADPH are both produced in the light-dependent reactions; *[1]*

Both molecules are required for reactions in the Calvin cycle *[1]*

4 102 kg *[2]*

5 Photosynthesis only occurs in the light; *[1]*

the rate of production is insufficient to supply the plant with the concentration of ATP required; *[1]*

some plant cells lack chloroplasts and would be unable to generate ATP *[1]*

17.4

1 (*Right of the line*)

RuBP accumulates because it has less CO_2 with which to react; *[1]*

GP concentration decreases because less RuBP and CO_2 are reacting together *[1]*

(*Left of the line*)

RuBP and GP are reacting at the same rate as they are being produced; *[1] [max 2]*

2 Temperature affects the rate of enzyme activity; *[1]*

more enzymes are involved in the light-independent stage than the light-dependent stage *[1]*

3 (*At low light intensity*)

GP accumulates because a lack of ATP and NADPH are produced in the light-dependent stage; *[1]*

RuBP and TP concentrations decrease because GP is not converted to TP; *[1]*

TP is not converted to RuBP; *[1]*

ATP and NADPH are required to convert GP to TP; *[1]*

ATP is required to convert TP to RuBP *[1]*

At high light intensity substances are reacting at the same rate as they are being produced *[1] [max 3]*

18.1

1 2 ATP used to convert glucose to hexose 1,6-bisphosphate; *[1]*

4 ATP produced in converting TP to pyruvate; *[1]*

Net of 2 ATP produced *[1]*

2 7 ATP per glucose; *[1]*

2 ATP from substrate-level phosphorylation; *[1]*

2 reduced NAD produced in glycolysis; *[1]*

which result in a maximum of 5 ATP; *[1]*

in oxidative phosphorylation *[1] [max 4]*

3 ATP phosphorylates glucose; *[1]*

to prepare it for subsequent reactions / to be broken down; *[1]*

TP is phosphorylated; *[1]*

by inorganic phosphate; [1]

ADP is phosphorylated when TP is converted to pyruvate [1] [max 4]

18.2

1 Enters mitochondria; [1]

(through) active transport; [1]

converted to acetyl groups; [1]

decarboxylated / loses carbon; [1]

dehydrogenated / loses hydrogen [1] [max 4]

2 Leaves the plant through stomata; [1]

(or) used in photosynthesis [1]

3 Subsequent stages of aerobic respiration take place in mitochondria; [1]

Enzymes of the Krebs cycle are located in mitochondrial matrix; [1]

Electron transport chains are located on inner mitochondrial membrane [1] [max 2]

18.3

1 4 [1]

2 10 ATP molecules (per turn); [1]

1 ATP from substrate-level phosphorylation; [1]

7.5 ATP from oxidative phosphorylation via reduced NAD; [1]

1.5 ATP via FAD [1]

3 *Idea of* rapid turnover / constantly reacting; [1]

reacts with acetyl CoA to form citrate [1]

18.4

Go further: How many ATP molecules are produced per glucose molecule?

1 Reduced NAD must be transported from the cytoplasm into the matrix. This requires ATP, and the reduced NAD may enter the ETC at a different point.

2 Reduced NAD enters at carrier 1 (because it is estimated to be responsible for 2.5 ATP molecules, which = 10 H$^+$ ions (from the four pumps) / 4

Reduced FAD enters at carrier 2 (because it is estimated to be responsible for 1.5 ATP molecules, which = 6 H$^+$ ions (from the final two pumps) / 4

Summary questions

1 Cristae are folds of the inner membrane; [1]

which increase the surface area available for oxidative phosphorylation [1]

2 Oxygen accepts electrons that have passed through the ETC; [1]

forming water; [1]

when combined with H$^+$ ions [1]

3 Some ATP is used for active transport of pyruvate (from the cytoplasm); [1]

ETC is inefficient / some energy is released as heat (and is not used to transport H$^+$ ions); [1]

some reduced coenzymes may be used in other processes (not the ETC) [1] [max 2]

18.5

1 Relatively small amounts of ATP are generated; [1]

insufficient chemical energy to sustain the functions of mammalian bodies [1]

2 Aerobic = 34.0% (((32 × 30.6) / 2880) × 100); [1]

anaerobic = 2.1% (((2 × 30.6) / 2880) × 100) [1]

3 a *Idea of* no terminal electron acceptor to accept electrons from the final electron carrier protein complex; [1]

proton gradient is disrupted because oxygen is not present to remove H$^+$ ions diffusing through ATP synthase [1]

b NAD is not regenerated because the ETC stops; [1]

the reactions needed to convert citrate back to oxaloacetate cannot occur [1]

18.6

1 Amine group; [1]

deamination [1]

2 $C_{15}H_{31}COOH + 23O_2 \rightarrow$ [1]

$16CO_2 + 16H_2O$ [1]

RQ = 16/23 = 0.696 [1]

3 a Oxaloacetate; [1]

both have 4 C atoms [1]

b Pyruvate; [1]

both have 3 C atoms [1]

19.1

Go further: The FTO 'hunger' gene: an example of the subtle effects of gene mutation?

1 Point mutation / substitution. The gene variants all produce functional proteins, albeit differing in their effect. Insertion and deletion mutations tend to have a more severe effect on phenotypes.

2 It might regulate the transcription of the ghrelin 'hunger hormone' (i.e. influences gene expression).

Summary questions

1 (The genetic code is a) triplet code; [1]

Three nucleotides represent a codon / code for one amino acid; [1]

Other codons in the DNA sequence will not be altered [1]

2 a 2 *[1]*

  b 1 *[1]*

  c 4 *[1]*

3 3 substitution mutations:

A to T (10th base); *[1]*

A to T (12th base); *[1]*

C to T (22nd base); *[1]*

deletion of 8 nucleotides (GATTATGG) *[1]*

19.2

1 Introns are removed from pre-mRNA; *[1]*

the remaining exons are spliced together to form mature mRNA *[1]*

2 Alternative splicing enables many different proteins to be translated from a single gene; *[1]*

this increases genetic diversity, which allows greater complexity in organisms *[1]*

3 Identical twins have 100% of the same DNA and fraternal twins have approximately 50% of the same DNA; *[1]*

there is a greater genetic influence on height than the risk of strokes; *[1]*

the evidence for this is that more than 90% of identical twins have the same height, whereas approximately 15% of identical twins share the risk of having a stroke *[1]*

19.3

1 The formation of connections between neurones; *[1]*

destruction of harmful immune cells; *[1]*

forming the shapes of organs and tissues; *[1]*

removing excess cells (e.g. from between digits); *[1]*

any other valid suggestion *[1] [max 2]*

2 Species differ in the complexity of their anatomy; *[1]*

the number of homeobox genes increases with complexity *[1]*

3 During necrosis, cells rupture and release enzymes; *[1]*

necrosis is an uncontrolled process; *[1]*

apoptosis is a controlled, regulated process; *[1]*

rather than cells rupturing, cell fragments are packaged into vesicles *[1]*

20.1

1 a Discontinuous; *[1]*

  b continuous; *[1]*

  c continuous *[1]*

2 Genotypes / genetics are identical (in clones); *[1]*

environmental differences (in temperature); *[1]*

the fruit grown at the higher temperature will be larger; *[1]*

due to greater enzyme activity; *[1]*

fruit size shows continuous variation *[1] [max 3]*

3 Genetics are controlled / identical in twins; *[1]*

the relative influence of genetics and the environment can be analysed; *[1]*

by monitoring differences that develop between the twins; *[1]*

idea that a trait that shows little variation between twins is likely to be largely determined by genetics (or reverse argument) *[1] [max 3]*

20.2

1 X chromosome is larger; *[1]*

X chromosome contains more genes than the Y chromosome *[1]*

2 $C^R C^R$ *[1]*

$C^R C^W$ *[1]*

3 a 0% *[1]*

  b 50% *[1]*

  c 12.5% *[1]*

  d 50% *[1]*

20.3

1 RR *[1]*

WW *[1]*

2 Round, yellow = 12/16 (75%) *[1]*

Round, green = 4/16 (25%) *[1]*

3 The genes are not linked / not on the same chromosome; *[1]*

the genes do not interact / there is no epistasis *[1]*

20.4

1 (Autosomal) linkage; *[1]*

genes on the same chromosome; *[1]*

allele combinations inherited together *[1] [max 2]*

2 Chi-squared value = 1.00; *[1]*

greater than 5% probability that the differences between observed and expected results are due to chance; *[1]*

scientist's predictions are supported / no significant difference between observed and expected phenotypes; *[1]*

genotypes are RrGG and rrGG (or rrGg) *[1]*

3 a (Epistatic gene produces) enzyme; *[1]*

homozygous recessive genotype results in no enzyme production; *[1]*

precursor molecule not converted *[1] [max 2]*

b (Epistatic gene produces) inhibitor / suppressor protein; [1]

Idea of modifies the other gene product; [1]

affects / AW transcription [1] [max 2]

20.5

1 Decrease in population size; [1]

some alleles are lost from the population [1]

2 $q = 0.008$ (0.007 669); [1]

$q^2 = 0.0000588$ [1]

3 13.1% are carriers [1]

$q = 0.0707$; [1]

$p = 0.929\ 29$; [1]

$2pq = 0.131\ 42$; [1]

0.8% of the general population are carriers, which is 12.3% less [1]

20.6

1 **a** Geographical isolation; [1]

b mechanical/anatomical isolation [1]

2 Sympatric speciation; [1]

behavioural isolation; [1]

gene flow restricted between the two groups [1] [max 2]

3 No: The genomes/DNA sequences of the two species are sufficiently different to consider them separate species; [1]

Yes: geographical isolation has caused their phenotypes to diverge, but not yet to the point where they are unable to breed together. [1]

21.1

1 In both procedures, molecules are separated by size; [1]

some molecules are slowed (by the stationary phase in chromatography) more than others [1]

2 **a** $10^{1.50}$ [1]

b $10^{3.61}$ [1]

c $10^{5.12}$ [1]

3 *Similarities:*

Each new DNA molecule consists of one old (template) strand and one new strand; [1]

free complementary nucleotides are joined to a template strand; [1]

Differences:

Only short fragments are replicated in PCR, whereas entire chromosomes are replicated naturally; [1]

PCR requires primers to be used; [1]

the DNA helicase enzyme separates strands in nature, whereas temperature cycling controls the process in PCR [1] [max 4]

21.2/21.3

Go further: Advances in sequencing

1 Cost, the length of DNA strands that can be sequenced during a single run, accuracy, the speed of sequencing.

Summary questions

1 **a** Particular allele base sequences are associated with certain diseases; [1]

People can be screened for the base sequences [1]

b Particular base sequences (barcodes) are unique to species; [1]

Similarities and differences in base sequences indicate the relatedness of different species [1]

2 *Similarities:*

Both use DNA polymerase; [1]

Both use free nucleotides [1]

Differences:

PCR for sequencing includes terminator bases; [1]

Different lengths of DNA are produced when sequencing [1]

Amplification uses a thermocycler, but sequencing uses capillaries / slides [1] [max 4]

3 mRNA can be modified after transcription; [1]

RNA splicing/ introns removed from RNA; [1]

Proteins can be modified after translation [1]

21.4

1 Restriction enzymes cut at specific recognition sites; [1]

the gene and the plasmid must have complementary sticky ends to be able to join together [1]

2 The herbicide will kill only weeds that are competing with the crop for resources; [1]

crop yield will increase and food prices will remain low [1]

3 The gene being studied is inactivated in embryonic stem cells (of mice); [1]

all cells in the study animal have the inactivated gene; [1]

example of gene that could be studied (e.g. tumour suppressor gene or proto-oncogene); [1]

the phenotypes of the knockout mice and normal mice are compared; [1]

the incidence of cancer is recorded [1] [max 2]

21.5

1 Somatic cell therapy is used; *[1]*

   this treats only the affected somatic tissue and the new alleles are not passed on to future generations; *[1]*

   the introduced alleles are not copied in mitosis; *[1]*

   the treated cells are replaced, meaning additional treatment is required *[1] [max 2]*

2 Delivering even a single gene via a vector is difficult; *[1]*

   delivering more than one gene would require a vector large enough to carry even more DNA; *[1]*

   the probability of all the new genes being integrated and functioning would be very low *[1] [max 2]*

3 The use of viruses as vectors taps into and adapts their natural mode of infection (i.e. inserting their DNA into specific cells); *[1]*

   liposomes are hydrophobic and are able to move through cell membranes *[1]*

22.1/22.2

1 (Meristem cells are) stem cells; *[1]*

   (Meristem cells are) totipotent; *[1]*

   (Meristem cells can) differentiate into a whole plant; *[1]*

   (by) vegetative propagation; *[1] [3 max]*

2 Auxin promotes root growth; *[1]*

   Enables differentiation of callus cells; *[1]*

   Enables plantlet formation; *[1] [2 max]*

3 Clones/new plants are genetically identical to the parent plant (i.e. no genetic variation); *[1]*

   Clones/new plants are equally susceptible to the pathogen; *[1]*

   The pathogen may be systemic/remains in the clones *[1] [2 max]*

22.3

Go further: Aphids

1 Reproduction is rapid (i.e. a faster rate than sexual reproduction would provide) and enables aphid populations to exploit beneficial conditions.

2 Eggs are more likely than nymphs to survive the cold temperatures of winter. Sexual reproduction also introduces genetic variation.

Summary questions

1 Reproductive cloning results in cloned whole organisms; *[1]*

   Non-reproductive cloning results in cloned cells/tissues/organs (but not whole organisms) *[1]*

2 Cells early in an embryo's development are totipotent; *[1]*

   These cells can develop into a whole organism; *[1]*

   Cells at a later stage are pluripotent/multipotent/not totipotent *[1][2 max]*

3 (Yes)

   No nuclear DNA will be present from the egg donor; *[1]*

   But mitochondrial DNA is transferred with the egg *[1]*

22.4/22.5

1 A process carried out by a microorganism is exploited for commercial benefit *[1]*

2 Microorganisms consume nutrients from the waste water; *[1]*

   (Organic) waste matter is therefore removed from the water *[1]*

3 Antibiotic chemicals are secreted by *Penicillium*; *[1]*

   The antibiotics are toxic to other microorganisms; *[1]*

   Food sources are preserved for *Penicillium* *[1] [2 max]*

22.6/22.7

1 Lack of nutrients; *[1]*

   Lack of oxygen; *[1]*

   Temperature that is too high or low; *[1]*

   Build-up of toxic waste products; *[1]*

   Incorrect pH *[1] [3 max]*

2 Population is allowed to reach stationary phase; *[1]*

   Secondary metabolites are produced; *[1]*

   Penicillin is a secondary metabolite; *[1]*

   Penicillin is produced as a defence mechanism *[1] [max 3]*

3 2.8×10^5 *[4]*

   If the final answer is incorrect, marks can be scored (up to a maximum of 3) for the following:

   Dilution factor of 20 000; *[1]*

   Estimate from plate 1 = 320 000; *[1]*

   Estimate from plate 2 = 240 000; *[1]*

   Mean = 280 000; *[1]*

22.8

1 Within a polysaccharide/cellulose/gelatin matrix; *[1]*

   One side of a semi-permeable membrane *[1]*

2 Set-up tends to be more expensive than the use of enzymes free in solution; *[1]*

   Less downstream processing, which saves money; *[1]*

   Continuous use increases efficiency and production rate; *[1]*

The balance of costs and benefits may depend on the type of reaction; [1]

For established biotechnology uses, the economic benefits outweigh the costs [1] [3 max]

3 Semi-synthetic penicillin has a slightly different structure to natural penicillin; [1]

Less bacterial resistance to semi-synthetic penicillin [1]

23.1/23.2

1 Inedible / indigestible parts; [1]

Energy used in respiration / to produce heat; [1]

Energy used for movement; [1]

Some energy lost in excretion [1] [max 3]

2 Restrict movement; [1]

Use antibiotics to reduce the amount of energy used by the immune system; [1]

Maintain optimal environmental temperature; [1]

More energy is converted to biomass [1] [max 3]

3 a 0.1 kg m^{-2} yr^{-1} [2]

b (Genetic modification of the crop could increase) resistance to disease/pests; [1]

Herbicide resistance; [1]

Drought resistance; [1]

This enables increased growth and higher yields [1] [max 3]

23.3

1 Nitrogen fixation; [1]

ammonification; [1]

nitrite; [1]

denitrification [1]

2 Deforestation; [1]

reduces photosynthesis; [1]

More combustion; [1]

in industry/vehicles/as fuel; [1]

Higher carbon dioxide concentrations in the atmosphere [1] [max 3]

3 Anaerobic conditions; [1]

Increased denitrification; [1]

Reduced nitrification; [1]

Lower concentration of nitrates for crop plants to absorb [1] [max 3]

23.4

1 (A plagioclimax is) a community resulting from deflected succession, where human influence has prevented succession from producing a climax community; [1]

(Examples include) managed forests, grazing grassland, arable crop fields [1]

2 Tolerance of extreme conditions/environments; [1]

The ability to fix nitrogen from the atmosphere; [1]

The ability to photosynthesise; [1]

The production of many seeds/spores that can be carried by wind; [1]

Rapid germination of seeds [1][max 3]

3 Primary succession is the natural development of an ecosystem to form a climax community; [1]

Deflected succession is when human activity interrupts succession, producing a plagioclimax; [1]

Primary succession tends to produce greater biodiversity because deflected succession reduces the number of ecological niches available [1]

23.5

1 96 000 [2]

[If the final answer is incorrect, '1 000 000 / 500' scores 1 mark]

2 Different areas within an ecosystem are identified; [1]

These areas vary in their abiotic conditions; [1]

Random sampling may miss some of the different areas [1] [max 2]

3 43 [2] [1 mark for 42.66]

24.1

1 a A factor that limits the maximum size of a population; [1]

b A factor that affects population size, but its effects vary with population density [1]

2 Birth rate is always higher than death rate; [1]

An increase in population size, with a relatively constant rate of change for much of the two stages; [1]

The increase occurs because death rate declines at an earlier stage than birth rate; [1]

This means the difference in the two rates widens during these stages [1] [max 3]

3 Country A has the faster population growth; [1]

A greater proportion of its population can be found in the younger age categories [1]

24.2/24.3

Go further: Allelopathy – competition between plants

1 When leaves are eventually shed and decompose, the allelopathic chemicals will diffuse through the soil.

2 Allelopathic chemicals may be persistent and remain in the soil for a long time, therefore interfering with the growth of crop plants.

Summary questions

1 Light; *[1]*

Water (from soil); *[1]*

Carbon dioxide; *[1]*

(Named) mineral ions (from soil) *[1] [3 max]*

2 The filter feeders have the same ecological niche; *[1]*

They feed on similar food items; *[1]*

Barnacles avoid competition; *[1]*

The competition is interspecific *[1] [3 max]*

3 (The predator peaks are lower because) there are fewer predators in the ecosystem; *[1]*

(The peaks are delayed because) when prey numbers rise, the chance of survival increases for predators; *[1]*

Intraspecific competition is reduced; *[1]*

After a short time, predator numbers rise; *[1]*

Reproduction and the rearing of offspring contribute to the delay *[1] [max 3]*

24.4/24.5

1 Conservation is active/manages an ecosystem/requires interference; *[1]*

Preservation avoids interference/aims to leave an ecosystem undisturbed *[1]*

2 Tilapia are lower down the food chain; *[1]*

Only a single energy transfer occurs (from producer to primary consumer); *[1]*

Farming salmon requires the depletion of wild (primary/secondary consumer) fish populations to act as food for the salmon *[1] [2 max]*

3 Sometimes it is unclear which species were present in the original community; *[1]*

Understanding and restoring the original abiotic conditions may be difficult; *[1]*

Succession takes a long time *[1][2 max]*

24.6–24.9

1 An ecosystem that is especially vulnerable to environmental change/human interference *[1]*

2 Human populations are no longer nomadic; *[1]*

The same land is grazed/deforested for long periods of time; *[1]*

Vegetation is unable to recover; *[1]*

The risk of soil erosion increases *[1] [3 max]*

3 Preservation is usually impossible because so few peat bogs remain in their original state; *[1]*

Peat bogs take a long time to form; *[1]*

There is a conflict between reclamation/conservation and human activity (e.g. extraction of peat) *[1]*

Answers to practice questions

Chapter 13

1 D *[1]*, **2** A *[1]*, **3** D *[1]*, **4** D *[1]*,
5 C *[1]*

6 **Level 3 [5–6 marks]:** Detailed comparison of the function and physiology of the two branches of the nervous system. There is a well-developed line of reasoning which is clear and logically structured. The information presented is relevant and substantiated.

Level 2 [3–4 marks]: Includes some correct comparisons of the two branches of the nervous system. There is a line of reasoning presented with some structure. The information presented is in the most part relevant and supported by some evidence.

Level 1 [1–2 marks]: Simple comments about the two branches. The information is basic and communicated in an unstructured way. The information is supported by limited evidence.

Relevant scientific points include:

Differences in neurotransmitters, ganglion position, general function (sympathetic = 'flight, fright, fight, or excite'; parasympathetic = 'rest and digest'), and examples of contrasting specific functions (e.g. sympathetic speeds up heart rate whereas parasympathetic slows heart rate).

Chapter 14

1 B *[1]*, **2** C *[1]*, **3** C *[1]*

4 **Level 3 [5–6 marks]:** Detailed description of both hormonal and nervous control of heart rate, with use of appropriate terminology. There is a well-developed line of reasoning which is clear and logically structured. The information presented is relevant and substantiated.

Level 2 [3–4 marks]: Accurate description of aspects of hormonal and nervous control. There is a line of reasoning presented with some structure. The information presented is in the most part relevant and supported by some evidence.

Level 1 [1–2 marks]: Simple comments about heart rate control. The information is basic and communicated in an unstructured way. The information is supported by limited evidence.

Relevant scientific points include:

adrenaline increases heart rate / stroke volume / cardiac output;

cardiovascular centre in medulla oblongata;

idea of nervous connection to SAN / sino-atrial node;

(which) controls frequency of waves of excitation / depolarisation;

vagus / parasympathetic , nerve decreases heart rate;

accelerator / sympathetic , nerve increases heart rate;

high blood pressure detected by stretch receptors / baroreceptors;

low blood pH / increased levels of blood CO_2 detected by chemoreceptors;

(receptors) in , aorta / carotid sinus / carotid arteries.

5 (Adrenaline is a) first messenger;

Binds to receptors on cell surface membranes;

cAMP is a second messenger;

(triggers intracellular) cascade effect;

Glycogenolysis;

Glucose is produced and diffuses out of cells   *[max 5]*

6 *[3]*

Trait	Type 1	Type 2
Treatment	Insulin injections	Dietary changes
Extent of genetic influence	Large / AW	Some / AW
Is insulin produced?	No / very little	Usually (but often less than normal)

Chapter 15

1 A *[1]*, **2** B *[1]*, **3** D *[1]*
4 A *[1]*, **5** A *[1]*, **6** C *[1]*

Chapter 16

1 D *[1]*, **2** A *[1]*

3 Some amylase is produced without gibberellin exposure;

Positive correlation;

Between gibberellin exposure and amylase production;

Amylase production plateaus after 13 days of gibberellin exposure;

Gibberellin triggers amylase production;

By switching on the gene coding for amylase

[max 5]

4 Auxin / one plant hormone can produce several effects depending on concentration;

Opposite effects can be produced by altering concentration;

Precise measurements of concentration are needed for tissue culture;

Concentrations will vary depending on the species;

Concentrations will vary depending on the stage of the culture   *[max 2]*

Chapter 17

1 D *[1],*   **2** D *[1],*   **3** A *[1]*

4 **Level 3 [5–6 marks]:** Detailed and accurate descriptions of the methods that can be used to improve primary productivity, with examples to illustrate each method. There is a well-developed line of reasoning which is clear and logically structured. The information presented is relevant and substantiated.

Level 2 [3–4 marks]: Includes some accurate descriptions of the methods used to manipulate productivity. There is a line of reasoning presented with some structure. The information presented is in the most part relevant and supported by some evidence.

Level 1 [1–2 marks]: Simple comments about the methods used on farms to improve productivity, with few correct examples. The information is basic and communicated in an unstructured way. The information is supported by limited evidence.

Relevant scientific points include:

Consideration of light banks, sowing timing, sowing density, greenhouse temperature regulation, irrigation, crop rotation, fertilisers, herbicides, pesticides.

5 Initially the increase in the rate of photosynthesis is proportional to the increase in light intensity (i.e. linear relationship);

At this point, light intensity is a limiting factor;

The graph plateaus at a high light intensity;

At this point, CO_2 concentration is a limiting factor
[max 3]

Chapter 18

1 C *[1],*   **2** B *[1],*   **3** D *[1],*   **4** B *[1]*

5 **Level 3 [5–6 marks]:** Detailed and accurate descriptions of the use of respirometers and the design of an experiment to measure RQ. There is a well-developed line of reasoning which is clear and logically structured. The information presented is relevant and substantiated.

Level 2 [3–4 marks]: Includes some accurate descriptions of the use of respirometers and valid experimental design. There is a line of reasoning presented with some structure. The information presented is in the most part relevant and supported by some evidence.

Level 1 [1–2 marks]: Simple comments about the use of respirometers and experimental design. The information is basic and communicated in an unstructured way. The information is supported by limited evidence.

Relevant scientific points include:

Details of the respirometer, including the role of soda lime, potassium hydroxide, the double chamber design, index liquid.

Details of a valid experimental design, including the use of a control respirometer with glass beads, control variables, measurement of the meniscus position, repeat readings.

An understanding of RQ calculation.

6 In respiration, small amounts of energy are released;

In a series of chemical reactions;

This enables the transfer of chemical energy to bonds in ATP;

In combustion, energy is released rapidly/in one go
[max 3]

Chapter 19

1 A *[1]*

2 **a** Lac operon *[1]*

b Lactose binds to the repressor protein;

Alters the shape of the repressor protein;

Prevents the repressor protein binding to the operator   *[max 2]*

3 Programmed;

Enzymes;

Blebs;

Phagocytes   *[max 4]*

4 Post-transcriptional/pre-translational (control);

Splicing;

Variation in which introns are retained and removed can produce different polypeptides;

Exons can be rearranged (to produce different polypeptides);   *[max 3]*

5 A = Substitution   B = Silent

C = Missense   *[max 3]*

Chapter 20

1 B *[1],*   **2** C *[1],*   **3** B *[1],*   **4** D *[1]*

5 C *[1]*

6 **Level 3 [5–6 marks]:** Accurate explanation of the founder effect with at least two detailed examples from populations. There is a well-developed line of reasoning, which is clear and logically structured. The information presented is relevant and substantiated.

Level 2 [3–4 marks]: Includes a description of the founder effect and one example from populations. There is a line of reasoning presented with some structure. The information presented is in the most part relevant and supported by some evidence.

Level 1 [1–2 marks]: Simple comments about the founder effect. The information is basic and communicated in an unstructured way. The information is supported by limited evidence.

Relevant scientific points include:

Genetic bottlenecks, reduced genetic diversity, examples such as Ellis–van Creveld syndrome, blood group distribution.

Chapter 21

1 B *[1]*,   2 C *[1]*,   3 D *[1]*

4 **A** DNA polymerase / Taq polymerase

   B Restriction endonucleases

   C (DNA) ligase

   D Plasmid(s)

   E Reverse transcriptase   *[max 5 marks]*

5 Genetic engineering = C

   Somatic cell therapy = D

   DNA profiling = B

   Animal reproductive cloning = A   *[max 4 marks]*

Chapter 22

1 Propagation; *[1]*   Meristem; *[1]*

   Suckers; *[1]*   Genetically; *[1]*

2 D, B, C, A   *[4]*

3 Finite / limited glucose supply;   *[1]*

   Glucose used up during the log phase;   *[1]*

   Penicillin produced in stationary phase;   *[1]*

   Secondary metabolites produced in times of stress / not produced as part of normal metabolism   *[1]*

4 (From left to right)

   Adsorption (to inorganic material / cellulose / silica); matrix entrapment; covalent or ionic bonding to inorganic carrier; membrane entrapment / encapsulation   *[max 4]*

Chapter 23

1 D *[1]*

2 **a** Oak (leaves) and nettle (leaves)   *[1]*

   b Sparrowhawk, badger, owl, blue tit   *[1]*

   c Blue tit population would reduce/disappear;

      Aphid population would rise;

      Nettle population would be reduced;

      Sparrowhawk population would be reduced;

      Owl population would be reduced;

      Blackbird population may reduce (as sparrowhawks have fewer blue tits on which to predate);

      Shrew population may reduce (as owls have fewer blue tits on which to predate);

      Badger population may decline   *[max 4]*

   d Shrew to badger / shrew to owl / blackbird to sparrowhawk / blue tit to sparrowhawk;

      More energy released as heat / more (named) indigestible parts   *[2]*

3 **a** 4% *[1]*   **b** 20% *[1]*   **c** 15% *[1]*

4 Nitrification = A, B, D   *[1]*

   Denitrification = C, G   *[1]*

   Nitrogen fixation = E, F   *[1]*

Chapter 24

1 C *[1]*

2 **Level 3 [5–6 marks]:** Detailed descriptions of abiotic changes, with clear links to the subsequent effects on biodiversity and the use of examples. There is a well-developed line of reasoning which is clear and logically structured. The information presented is relevant and substantiated.

   Level 2 [3–4 marks]: Includes some accurate descriptions of abiotic changes, with some links to the effects on biodiversity. There is a line of reasoning presented with some structure. The information presented is in the most part relevant and supported by some evidence.

   Level 1 [1–2 marks]: Simple comments about abiotic changes or biotic changes, with few correct examples. The information is basic and communicated in an unstructured way. The information is supported by limited evidence.

   Relevant scientific points include:

   Effects on climate (e.g. climate change), soils (e.g. salinity changes), and water quality (e.g. eutrophication).

   Impact of climate change on vulnerable species (e.g. frog species), deforestation causing extinction/ reduced biodiversity, the impact of pesticides on some species.

3 Stage 1: resources are plentiful and population size increases;   *[1]*

   Stage 2: Resources become limited and population begins to decrease;   *[1]*

   Stage 3: Competition is reduced and the population rises again;   *[1]*

4 Stage 1: Prey population rises, which provides more food for predators;   *[1]*

   Stage 2: Prey population declines because of increased predation;   *[1]*

   Stage 3: Predator population decreases because of increased intraspecific competition for reduced prey numbers;   *[1]*

   Stage 4: Prey population rises again because predator numbers have decreased.   *[1]*

Appendix (Statistics data tables)

▼ *Table of values of* t

Degree of freedom (df)	p values			
	0.10	0.05	0.01	0.001
1	6.31	12.71	63.66	636.60
2	2.92	4.30	9.92	31.60
3	2.35	3.18	5.84	12.92
4	2.13	2.78	4.60	8.61
5	2.02	2.57	4.03	6.87
6	1.94	2.45	3.71	5.96
7	1.89	2.36	3.50	5.41
8	1.86	2.31	3.36	5.04
9	1.83	2.26	3.25	4.78
10	1.81	2.23	3.17	4.59
12	1.78	2.18	3.05	4.32
14	1.76	2.15	2.98	4.14
16	1.75	2.12	2.92	4.02
18	1.73	2.10	2.88	3.92
20	1.72	2.09	2.85	3.85
α	1.64	1.96	2.58	3.29

▼ *Critical values for Spearman's rank correlation coefficient,* r_s

	$p = 0.1$	$p = 0.05$	$p = 0.02$	$p = 0.01$			$p = 0.1$	$p = 0.05$	$p = 0.02$	$p = 0.01$
	5%	$2\frac{1}{2}$%	1%	$\frac{1}{2}$%	1-Tail Test		5%	$2\frac{1}{2}$%	1%	$\frac{1}{2}$%
	10%	5%	2%	1%	2-Tail Test		10%	5%	2%	1%
n						n				
1	–	–	–	–		31	0.3012	0.3560	0.4185	0.4593
2	–	–	–	–		32	0.2962	0.3504	0.4117	0.4523
3	–	–	–	–		33	0.2914	0.3449	0.4054	0.4455
4	1.0000	–	–	–		34	0.2871	0.3396	0.3995	0.4390
5	0.9000	1.0000	1.0000	–		35	0.2829	0.3347	0.3936	0.4328
6	0.8286	0.8857	0.9429	1.0000		36	0.2788	0.3300	0.3882	0.4268
7	0.7143	0.7857	0.8929	0.9286		37	0.2748	0.3253	0.3829	0.4211
8	0.6429	0.7381	0.8333	0.8810		38	0.2710	0.3209	0.3778	0.4155
9	0.6000	0.7000	0.7833	0.8333		39	0.2674	0.3168	0.3729	0.4103
10	0.5636	0.6485	0.7455	0.7939		40	0.2640	0.3128	0.3681	0.4051
11	0.5364	0.6182	0.7091	0.7545		41	0.2606	0.3087	0.3636	0.4002
12	0.5035	0.5874	0.6783	0.7273		42	0.2574	0.3051	0.3594	0.3955
13	0.4835	0.5604	0.6484	0.7033		43	0.2543	0.3014	0.3550	0.3908
14	0.4637	0.5385	0.6264	0.6791		44	0.2513	0.2978	0.3511	0.3865
15	0.4464	0.5214	0.6036	0.6536		45	0.2484	0.2945	0.3470	0.3822
16	0.4294	0.5029	0.5824	0.6353		46	0.2456	0.2913	0.3433	0.3781
17	0.4142	0.4877	0.5662	0.6176		47	0.2429	0.2880	0.3396	0.3741
18	0.4014	0.4716	0.5501	0.5996		48	0.2403	0.2850	0.3361	0.3702
19	0.3912	0.4596	0.5351	0.5842		49	0.2378	0.2820	0.3326	0.3664
20	0.3805	0.4466	0.5218	0.5699		50	0.2353	0.2791	0.3293	0.3628

▼ *continued*

	p = 0.1	p = 0.05	p = 0.02	p = 0.01
21	0.3701	0.4364	0.5091	0.5558
22	0.3608	0.4252	0.4975	0.5438
23	0.3528	0.4160	0.4862	0.5316
24	0.3443	0.4070	0.4757	0.5209
25	0.3369	0.3977	0.4662	0.5108
26	0.3306	0.3901	0.4571	0.5009
27	0.3242	0.3828	0.4487	0.4915
28	0.3180	0.3755	0.4401	0.4828
29	0.3118	0.3685	0.4325	0.4749
30	0.3063	0.3624	0.4251	0.4670

	p = 0.1	p = 0.05	p = 0.02	p = 0.01
51	0.2329	0.2764	0.3260	0.3592
52	0.2307	0.2736	0.3228	0.3558
53	0.2284	0.2710	0.3198	0.3524
54	0.2262	0.2685	0.3168	0.3492
55	0.2242	0.2659	0.3139	0.3460
56	0.2221	0.2636	0.3111	0.3429
57	0.2201	0.2612	0.3083	0.3400
58	0.2181	0.2589	0.3057	0.3370
59	0.2162	0.2567	0.3030	0.3342
60	0.2144	0.2545	0.3005	0.3314

▼ *Table of values of chi-squared*

df	p values								df
	0.99	0.95	0.90	0.50	0.10	0.05	0.01	0.001	
1	0.0001	0.0039	0.016	0.46	2.71	3.84	6.63	10.83	1
2	0.02	0.10	0.21	1.39	4.60	5.99	9.21	13.82	2
3	0.12	0.35	0.58	2.37	6.25	7.81	11.34	16.27	3
4	0.30	0.71	1.06	3.36	7.78	9.49	13.28	18.46	4
5	0.55	1.14	1.61	4.35	9.24	11.07	15.09	20.52	5
6	0.87	1.64	2.20	5.35	10.64	12.59	16.81	22.46	6
7	1.24	2.17	2.83	6.35	12.02	14.07	18.48	24.32	7
8	1.65	2.73	3.49	7.34	13.36	15.51	20.09	26.12	8
9	2.09	3.32	4.17	8.34	14.68	16.92	21.67	27.88	9
10	2.56	3.94	4.86	9.34	15.99	18.31	23.21	29.59	10
11	3.05	4.58	5.58	10.34	17.28	19.68	24.72	31.26	11
12	3.57	5.23	6.30	11.34	18.55	21.03	26.22	32.91	12
13	4.11	5.89	7.04	12.34	19.81	22.36	27.69	34.53	13
14	4.66	6.57	7.79	13.34	21.06	23.68	29.14	36.12	14
15	5.23	7.26	8.55	14.34	22.31	25.00	30.58	37.70	15
16	5.81	7.96	9.31	15.34	23.54	26.30	32.00	39.29	16
17	6.41	8.67	10.08	16.34	24.77	27.59	33.41	40.75	17
18	7.02	9.39	10.86	17.34	25.99	28.87	34.80	42.31	18
19	7.63	10.12	11.65	18.34	27.20	30.14	36.19	43.82	19
20	8.26	10.85	12.44	19.34	28.41	31.41	37.57	45.32	20
21	8.90	11.59	13.24	20.34	29.62	32.67	38.93	46.80	21
22	9.54	12.34	14.04	21.34	30.81	33.92	40.29	48.27	22
23	10.20	13.09	14.85	22.34	32.01	35.17	41.64	49.73	23
24	10.86	13.85	15.66	23.34	33.20	36.42	42.98	51.18	24
25	11.52	14.61	16.47	24.34	34.38	37.65	44.31	52.62	25
26	12.20	15.38	17.29	25.34	35.56	38.88	45.64	54.05	26
27	12.88	16.15	18.11	26.34	36.74	40.11	46.96	55.48	27
28	13.56	16.93	18.94	27.34	37.92	41.34	48.28	56.89	28
29	14.26	17.71	19.77	28.34	39.09	42.56	49.59	58.30	29
30	14.95	18.49	20.60	29.34	40.26	43.77	50.89	59.70	30
40	22.16	26.51	29.05	39.34	51.81	55.76	63.69	73.40	40
60	37.48	43.19	46.46	59.33	74.40	79.08	88.38	99.61	60
80	53.54	60.39	64.28	79.33	96.58	101.88	112.33	124.84	80
100	70.06	77.93	82.36	99.33	118.50	124.34	135.81	149.45	100